Intelligent Systems Reference Library

Volume 227

The aim of this series is to publish a Reference Library, including novel advances and developments in all aspects of Intelligent Systems in an easily accessible and well structured form. The series includes reference works, handbooks, compendia, textbooks, well-structured monographs, dictionaries, and encyclopedias. It contains well integrated knowledge and current information in the field of Intelligent Systems. The series covers the theory, applications, and design methods of Intelligent Systems. Virtually all disciplines such as engineering, computer science, avionics, business, e-commerce, environment, healthcare, physics and life science are included. The list of topics spans all the areas of modern intelligent systems such as: Ambient intelligence, Computational intelligence, Social intelligence, Computational neuroscience, Artificial life, Virtual society, Cognitive systems, DNA and immunity-based systems, e-Learning and teaching, Human-centred computing and Machine ethics, Intelligent control, Intelligent data analysis, Knowledge-based paradigms, Knowledge management, Intelligent agents, Intelligent decision making, Intelligent network security, Interactive entertainment, Learning paradigms, Recommender systems, Robotics and Mechatronics including human-machine teaming, Self-organizing and adaptive systems, Soft computing including Neural systems, Fuzzy systems, Evolutionary computing and the Fusion of these paradigms, Perception and Vision, Web intelligence and Multimedia.

Indexed by SCOPUS, DBLP, zbMATH, SCImago.

All books published in the series are submitted for consideration in Web of Science.

Larisa Ivascu · Lucian-Ionel Cioca ·
Florin Gheorghe Filip
Editors

Intelligent Techniques for Efficient Use of Valuable Resources

Knowledge and Cultural Resources

Springer

Editors
Larisa Ivascu
Faculty of Management in Production
and Transportation
Politehnica University of Timisoara
Timisoara, Romania

Lucian-Ionel Cioca
Faculty of Engineering
Lucian Blaga University of Sibiu
Sibiu, Romania

Florin Gheorghe Filip
The Romanian Academy
Bucharest, Romania

ISSN 1868-4394 ISSN 1868-4408 (electronic)
Intelligent Systems Reference Library
ISBN 978-3-031-09930-4 ISBN 978-3-031-09928-1 (eBook)
https://doi.org/10.1007/978-3-031-09928-1

This Springer imprint is published by the registered company Springer Nature Switzerland AG
The registered company address is: Gewerbestrasse 11, 6330 Cham, Switzerland

Preface

In the modern society, there is an ever-growing trend to use ever more resources. At the same time, a major progress in the development and usage of Information Technology (IT) can be noticed. IT is almost ubiquitous in our professional and personal life. Almost all areas of human activity innovate and use IT power with a view to adapt themselves to market requirements and human well-being. Since the available usable resources are limited, IT involvement must be adapted to the needs of sustainable development meant to avoid irrational and unsustainable behavior.

The development and usage of information technologies and mobile communications themselves can cause a nonnegligible consumption of resources, in particular, energy and cause an increase in the content of Carbon in the earth's atmosphere. Therefore, there are important efforts to create *green computing* tools and methods that are getting traction in academia and industry. Such trends and the corresponding results are presented in the literature.

At the same time, the usage of Information and communication technologies can be viewed as means meant to contribute to the rational usage of other resources. This volume aims to present a multidisciplinary view meant to illustrate several significant efforts and results about the contribution of information and communication technologies to make available new resources and rationally use the existing ones. Authors from various countries have been invited to contribute so that a rather broad and balanced image about the current trends and recent results obtained could result. The proposed chapters address IT and the efficient use of various resources such as manpower and human knowledge, natural resources including water, raw materials and end-of-life products, financial assets, data sets and computer software, and even cultural goods.

The volume is intended to be insightful for researchers, instructors, and planers from various domains and can also be used as an auxiliary material for postgraduate studies in applied informatics, business administration, industrial engineering, and digital humanities.

The volume is organized into ten chapters as follows.

Chapter 1—It is an introductory chapter that presents the importance of ICT for enterprises based on the principles of the circular economy (CE) and the characteristic features of Industry 5.0. The paper aims to identify the role of information and communications technology in the CE context. It also presents economic, social, and technical constraints and barriers that need to be addressed to pave the way for the implementation of the principles of the circular economy.

Chapter 2—The general objective of this chapter is to study the financial sustainability of a public university in terms of financial indicators. The first part of the research aimed at a scientific-theoretical basis that was based on qualitative research in order to thoroughly investigate the current state of knowledge. The qualitative part is completed by a quantitative study based on a mathematical model of the financial sustainability of Romanian universities.

Chapter 3—The chapter presents the concepts and technology of virtual exhibitions, with a particular emphasis on the changes brought on by the current COVID-19 pandemic Modern information and communication technologies are used to promote a new paradigm for facilitating the access to cultural heritage collections, by implementing virtual exhibitions, in the form of web applications, accessible online, as well as native mobile applications for mobile devices.

Chapter 4—It addresses *eTeaching* and *eLearning* as enabling means for teachers and students to take online courses with the help of electronic devices such as computers, laptops, tablets, and even smartphones. This chapter analyses the challenges that universities, professors, and students have had to overcome in order to build a reliable and efficient electronic educational system.

Chapter 5—The restrictive sanitary measures, taken worldwide, for the COVID-19 pandemic moved teaching activities into online environment. These circumstances have forced the restructuring of the face-to-face teaching process. This chapter describes, in its first part, the exploratory insights within the literature, and in the second part, with empirical evidence which highlights those students prefer the hybrid teaching system, and that teachers have managed to overcome the barriers/obstacles generated by the transition from traditional to the online teaching system.

Chapter 6—The chapter presents an integrated system concept, starting from the need for a rigorous approach regarding new healthcare product development. The chapter contains a theoretical part about machine learning in healthcare, *Agile* approach for healthcare organizations' projects and model based on the modified *Quality Function Deployment* (QFD) method *for Agile* applied for healthcare products.

Chapter 7—The chapter considers energy-efficient data centers both in four-season countries and in two-seasons countries. Apart from those data centers, another open problem is how to reduce energy use in P2P networks and briefly discusses such issues too. The presented ideas merely reflect a physicist's perspective on this subject.

Chapter 8—The chapter aims at revealing the relationship between the construct of technostress creators, including its five dimensions, and employee well-being in terms of work–life balance and job burnout. In doing this, quantitative data were collected in Lithuania.

Chapter 9—In this chapter, a methodological approach useful to identify the critical treatment phases and evaluate the effectiveness of the diverse upgrade interventions is proposed. Performance indicators useful for the characterization of the *Water Treatment Plants* (WTP), after a preliminary monitoring phase, are presented. In addition, examples of functionality tests applied to real WTPs are shown.

Chapter 10—This chapter validates and discusses the application of two intelligent learning control techniques, namely *Model-free Value Iteration Reinforcement Learning* (MFVIRL) and *Virtual State-feedback Reference Tuning* (VSFRT), for linear output reference model (ORM) tracking of three inexpensive lab scale systems which are interacted with by the help of modern software and hardware.

The editors of the volume are grateful to prof. L. Jain for his valuable guidance and support.

Timisoara, Romania — Larisa Ivascu
Sibiu, Romania — Lucian-Ionel Cioca
Bucharest, Romania — Florin Gheorghe Filip
March 2022

Contents

Editors and Contributors

About the Editors

Ivascu Larisa is a full professor from Politehnica University of Timisoara—Romania, Faculty of Management in Production and Transportation, Department of Management and has an overall experience of 20 years of programming, teaching, and research. In 2007, she graduated from the Faculty of Computer Science, specializing in Software at Politehnica University of Timisoara (UPT). In 2010 she graduated Master in Business Administration, from the Faculty of Management in Production and Transportation. She becomes Dr. Eng. (Ph.D. degree, 2013) of Politehnica University of Timisoara—Romania. She is the director of the Research Center in Engineering and Management and head of the Entrepreneurship Office of Politehnica University of Timișoara. She is the president of the scientific committee of the Academy of Political Leadership, vice-president of the Society for Ergonomics and Work Environment Management, and a member of the World Economics Association (WEA), International Economics Development and Research Center (IEDRC), and Academic Management Society of Romania. Since 2021, she is a member of the Advisory Board for Research and Development and Innovation—the Ministry of Research, Innovation and Digitalization, and a member of the Academy of Romanian Scientists. She is the director, manager, or member of national and international projects. She has been involved in projects with local authorities. She has published 9 books, contributed scientifically to over 190 scientific research, and was invited to participate in numerous international and national events. She is the guest editor of many prestigious journals. She has coordinated the volumes of prestigious publishing houses and forms research teams with researchers from all over the world. More details can be found at: http://www.mpt.upt.ro/eng/research/pdf/CCIM/CV_Ivascu%20Larisa_eng.pdf.

Cioca Lucian-Ionel is a full professor at the "Lucian Blaga" University of Sibiu, since 2007. Since 2010, he is a *doctoral advisor* in *Engineering and management*, and since 2012–present he is a member of the *National Council for the Certification of University Titles, Diplomas and Certificates*, the Commission for *Engineering and Production Management*, the Ministry of Education. Since 2017–present, he is a member of the *Specialized Certification Committee*, the Ministry of Education, and also since 2017, he is a member of *the Advisory Board for Research and Development and Innovation*, the Ministry of Research, Innovation and Digitalization. His research focuses on the following directions: management, human resources management, production systems engineering, ergonomics, circular economy, sustainability, occupational safety, and health management. He has published over 250 scientific papers, of which over 135 are indexed by Clarivate Analytics on the Web of Science. He is a professional member: The Academy of Romanian Scientists, The General Association of Engineers in Romania, The Association of Doctoral and Excellence Activities in Business Engineering and Management, The German General Engineer's Association, The World Economics Association (WEA), United Kingdom, The Romanian Managers and Economical Engineers Association, and The Transylvanian Association for the Literature and Culture of the Romanian People. He is also editor in chief, *editorial board member*, and guest editors of the several journals. He is the *Director* of the Master's training program *Occupational Safety, Health and Work Relations Management*, and the *Director* of the Postgraduate training and professional development program *Risk Assessment and Audit in Occupational Health and Safety*.

Filip Florin Gheorghe was born in 1947 in Bucharest, Romania. He graduated in *Control Engineering* at *Politehnica* University of Bucharest in 1970 and received his Ph.D. degree from the same university in 1982. He was elected as a corresponding member of the Romanian Academy in 1991 and became a full member of the Academy in 1999. During 2000–2010, he was vice-president of the Romanian Academy (elected in 2000, re-elected in 2004, and 2006). In 2010, he was elected president of the *Information Science and Technology* section of the Academy (re-elected in 2015, and 2019). He was the managing director of the *National Institute for R&D in Informatics-ICI*, Bucharest (1991–1997). He is an honorary member of the Romanian Academy of Technical Sciences and Academy of Sciences of Republic of Moldova. He was the chair of IFAC TC 5.4 (Large-scale Complex Systems) from 2002 to 2008. His main scientific interests include optimization and control of large-scale complex systems, decision support systems (DSS), technology management and foresight, and IT applications in the cultural sector. He authored/co-authored over 350 papers published in international journals (IFAC J Automatica, IFAC J Control Engineering Practice, Annual Reviews in Control, Computers in Industry, Large-Scale Systems, Technological and Economic Development of Economy, and so on) and contributed to volumes printed by international publishing houses (Pergamon Press, Elsevier, Kluwer, Chapman & Hall and so on). He is also the author/co-author of thirteen monographs (published in Romanian, English, and French by Editura Tehnică, Hermès-Lavoisier, Paris, J. Wiley & Sons, London, Springer) and editor/co-editor of 30 volumes of contributions (published by Editura Academiei Române, Pergamon Press, North Holland, Elsevier, IEEE Computer Society, and so on). He presented invited lectures in universities and research institutions, and plenary papers at scientific conferences in Brazil, Chile, China, France, Germany,

Lithuania, Poland, Portugal, the Republic of Moldova, Romania, Spain, Sweden, Tunisia, and the UK. More details can be found at: https://acad.ro/cv/FilipF/FGF-CV-en.pdf.

Contributors

Abbà Alessandro Department of Civil, Environmental, Architectural Engineering and Mathematics, University of Brescia, Brescia, Italy

Artene Alin Faculty of Management in Production and Transportation, Politehnica University of Timisoara, Timisoara, Romania

Bertanza Giorgio Department of Civil, Environmental, Architectural Engineering and Mathematics, University of Brescia, Brescia, Italy

Bobițan Nicolae Faculty of Economics and Business Administration, West University of Timisoara, Timisoara, Romania

Bogdan Oana Faculty of Economics and Business Administration, West University of Timisoara, Timisoara, Romania

Borlea Alexandra-Bianca Department of Automation and Applied Informatics, Politehnica University of Timisoara, Timisoara, Romania

Breaz Teodora Odett Faculty of Economic Sciences, "1 Decembrie 1918" University of Alba Iulia, Alba Iulia, Romania

Burcă Valentin Faculty of Economics and Business Administration, West University of Timisoara, Timisoara, Romania

Caccamo Francesca Maria Department of Civil and Architectural Engineering, University of Pavia, Pavia, Italy

Calatroni Silvia Department of Civil and Architectural Engineering, University of Pavia, Pavia, Italy

Carnevale Miino Marco Department of Civil and Architectural Engineering, University of Pavia, Pavia, Italy

Christianto Victor Malang Institute of Agriculture, Malang, Indonesia

Cioca Lucian-Ionel Faculty of Engineering, Lucian Blaga University of Sibiu, Sibiu, Romania

Ciurea Cristian Department of Economic Informatics and Cybernetics, Bucharest University of Economic Studies, Bucharest, Romania

Collivignarelli Maria Cristina Department of Civil and Architectural Engineering, University of Pavia, Pavia, Italy;
Interdepartmental Centre for Water Research, University of Pavia, Pavia, Italy

Domil Aura Emanuela Faculty of Economics and Business Administration, West University of Timisoara, Timisoara, Romania

Dumitrescu Diana Faculty of Economics and Business Administration, West University of Timisoara, Timisoara, Romania

Filip Florin Gheorghe The Romanian Academy, Bucharest, Romania

Fülöp Melinda Timea Faculty of Economics and Business Administration, Babeş-Bolyai University, Cluj-Napoca, Romania

Ionica Andreea Department of Management and Industrial Engineering, University of Petrosani, Petrosani, Romania

Ivascu Larisa Faculty of Management in Production and Transportation, Research Center for Engineering and Management, Politehnica University of Timisoara, Timisoara, Romania;
Academy of Romanian Scientists, Bucharest, Romania

Leba Monica Department of System Control and Computer Engineering, University of Petrosani, Petrosani, Romania

Radac Mircea-Bogdan Department of Automation and Applied Informatics, Politehnica University of Timisoara, Timisoara, Romania

Smarandache Florentin Mathematics, Physical and Natural Sciences Division, University of New Mexico, Gallup, NM, USA

Stankevičiūtė Živilė School of Economics and Business, Kaunas University of Technology, Kaunas, Lithuania

Umniyati Yunita Department of Mechatronics, Swiss-German University, Tangerang, Indonesia

Chapter 1
Information and Communications Technology in the Context of Circular Economy

Larisa Ivascu, Lucian-Ionel Cioca, and Florin Gheorghe Filip

Abstract Sustainability is being addressed by more and more organizations. In the context of sustainable global development, the circular economy is gaining interest from governments, people and various organizations. The circular economy has the main measure of reducing the amount of waste generated and of streamlining the number of organizational resources. In this context, companies are attracted to these concepts that protect the environment. Organizational interests are diverse, but the most important are the financial benefits. The paper aims to identify the role of information and communications technology in the CE context. It also presents economic, social, and technical constraints and barriers that need to be addressed in order to pave the way for the implementation of the principles of the circular economy.

Keywords Clean production · Environmental responsibility · ICT · Raw materials · Redesign · Reuse · SDG · Sustainability · Technological responsibility

1.1 Introduction

Sustainability is a concept realized and observed be an ever larger number of actants, governments, people, and companies. It contributes to achieving a balance between economic, social, and environmental responsibilities. At the same time, it stimulates the reduction of resource consumption to give future generations the opportunity to use the same resources as today [16, 23]. However, marketing has led to the creation

---

L. Ivascu
Faculty of Management in Production and Transportation, Research Center for Engineering and Management, Politehnica University of Timisoara, Timisoara, Romania
e-mail: larisa.ivascu@upt.ro

L.-I. Cioca (✉)
Faculty of Engineering, Lucian Blaga University of Sibiu, Sibiu, Romania
e-mail: lucian.cioca@ulbsibiu.ro

F. G. Filip
The Romanian Academy, Bucharest, Romania
e-mail: ffilip@acad.ro

L. Ivascu et al. (eds.), *Intelligent Techniques for Efficient Use of Valuable Resources*, Intelligent Systems Reference Library 227,
https://doi.org/10.1007/978-3-031-09928-1_1

of new needs and desires. In this way, consumers buy more and more goods, and organizations produce more and more [6, 21]. So, we are at the intersection of two concepts: *marketing* that creates needs and desires to increase demand on the market, and *sustainability* that recommends efficiency and reduction of consumption to create a balance.

The *circular economy* (CE) is a new pathway to sustainability. The circular economy involves a series of activities aiming at, among other things, reducing excessive consumption, eliminating waste, and regenerating the natural ecosystem. The CE concept is gaining more and more interest in environmental and economic responsibilities [2, 5]. At the same time, it is an approach found in the decision-making processes and in the thoughts of the interested parties to obtain a sustainable development of the organization [24, 26]. The approach is found in both, developing, and developed countries, as well, and is becoming more and more apparent [17]. The more and more growing preoccupations for the natural environment quality and limited resources as a part of sustainable development have led to concepts such as circular economy [28], environmentally conscious manufacturing and product recovery trends [14]. To make those activities effective, several multicriteria decision-making methods can be used [13].

CE can be viewed as an alternative to the linear economy model. In linear economics, raw materials are used to make a product that, when it reaches the end of its life cycle, is thrown away and becomes waste. In the CE, the flow focuses on recycling and reuse. Therefore, at the end of the life cycle of the product, it is recycled or reused to reduce the amount of waste to the extent possible [8]. The CE reduces pressure on the environment and contributes significantly to reducing the number of resources used to make products [4, 31].

The chapter begins with an overview of the importance of sustainability for organizations and the planet. Then the connections between CE, sustainability and SDGs are presented. In the second part of the chapter there is a presentation of the implications of ICT in the circular economy. The chapter concludes with proposals for the adoption of ICT for a sustainable approach to operational flow and organizational activity by proposing an architecture. This chapter aims to identify the role of information and communications technology in the CE context. It also presents economic, social, and technical constraints and barriers that need to be addressed to pave the way for the implementation of the principles of the circular economy.

1.2 Sustainability Functions and Their Importance

In the recent time, the involvement of companies in approaching sustainability has become ever more intense [25]. Even if there is no obligation to implement sustainability, organizations are beginning to set general sustainability goals that they can achieve [11, 15]. The setting these objectives is done in accordance with the vision, mission, and operational activity of an organization [31].

The approach of sustainability implies an efficient usage of the organizational resources to protect the environment and offer to the future generations the same development opportunities [9]. This approach involves an active involvement of the organization and the fulfillment of some functions of sustainability. These functions of sustainability also aim at the organizational implications in the circular economy. They are called "R" functions of sustainability and include *reimagining, reducing, reusing, redesigning, recycling, rethinking, repairing, reconditioning, remanufacturing, recovering, and rejecting* [1, 18, 27]. The implications of the 11 Rs functions are:

- *Reimaging*—building a sustainable and competitive image of the organization. The purpose of this function is to think creatively and develop new requirements for product design.
- *Reduction*—emphasis is placed on reducing the number of resources used in the operational flow. The purpose of this function is to create a stream of waste and minimal raw materials. This function can be applied in the stages: product design, product development, logistics chain, collaborator chain and promotion.
- *Reuse*—is a function that characterizes the companies that have developed potential strategies in this regard. The following actions are known for this dimension: reuse by lease, reuse by sale, reuse by repair, reuse by donation and reuse by loan.
- *Redesign*—is applied in the product design phase and aims to improve production conditions. The redesign starts from the material used and continues with the product sketches.
- *Recycling*—is one of the functions approached by more and more companies. It aims to recover the value of waste and reuse it in the operational flow.
- *Rethinking*—addresses stakeholders and aims at a new approach to organizational strategies and objectives. Actions are performed from the design stage of the product.
- *Repair*—refers to those functions or features that are no longer in line with the initial state. Actions are carried out with the aim of putting the product back into operation or bringing the condition of the product to a level of use.
- *Reconditioning*—aims at reducing actions as impact that may involve even partial dismantling. The purpose of the function is to put the product back into operation.
- *Remanufacturing*—unlike reconditioning, the actions are more numerous and can even lead to the total disassembly of the product. The purpose of the function is to put the product back into operation.
- *Recovery*—refers to the value of the product and those actions meant to lead to the recovery of parts of the product. The purpose of this step is to use and efficiently apply the materials in the operational flow.
- *Refuse (also called Disposal)*—some materials and raw materials should be rejected from the design stage of the product. These materials have an increased impact on the environment and should be discarded from the outset.

All these functions contribute to the increase of the organizational implications in the sustainable development and of the efficiency of the consumption of organizational resources. All these functions are developed in accordance with the 17 objectives of sustainable development and the 169 targets. The importance of these functions increases in proportion to the organizational involvement, the annual or biannual reporting of sustainability and the requirements of the partners. Many of the organizations approach these functions for the adoption of the circular economy in the operational flow.

1.3 Circular Economy: ICT, Principles, Barriers, and Drivers

The traditional economy does not support the recovery of materials and the functions of sustainability. The amount of waste has been so far continuously increased. It has been proven over time that the conservation of material and energy can be achieved at the organizational level only by adopting a loop-type system that minimizes waste. As shown in Fig. 1.1, the linear economy model neglects the value of the product when it reaches the end of its life cycle. Figure 1.2 shows the basic model of the circular economy in which the emphasis is on recycling and reusing materials for economic stimulation. In this model, the amount of waste is desired to be zero, reducing the negative impact on the environment and renewable resources. The major difference is at the end of a product's life cycle.

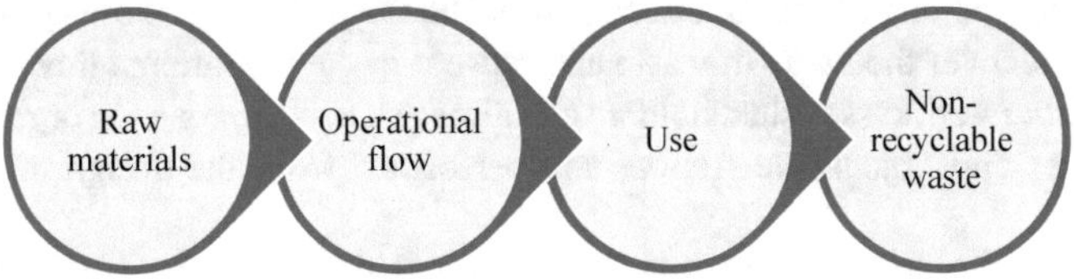

Fig. 1.1 Linear economy

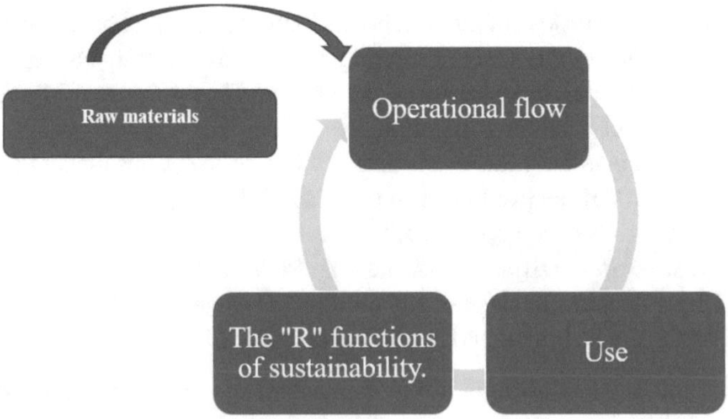

Fig. 1.2 Circular economy

Table 1.1 The principles of the circular economy

Input	Concept	Output
Raw materials =>	**Circular economy**	Zero waste
		Production Manufacturing
		Distribution
		Consumption
R function (redesign, reimage) =>		R function
		Sustainability
Residual waste (Disposal) <=		

Table 1.1 summarizes the operating principle of the circular economy by taking into account the inputs, outputs and R functions of sustainability. In this table, it can be seen that the input into the system is "raw materials" and some R functions of sustainability (redesign, reimagining, rejection and rethinking). The outputs of this system are: zero waste, production, manufacturing, distribution, Rs functions (reduction, reuse, recycling, repair, reconditioning, remanufacturing and recovery.)

As shown above, the circular economy contributes to achieving organizational goals. The *purpose* of the EC, at organizational level can be:

- reducing excessive resource consumption,
- reduces environmental footprint
- generate increased wealth
- reducing the amount of waste
- regeneration of the natural ecosystem
- regeneration of human capital
- optimizing resource consumption
- reduction of greenhouse gas emissions
- sustainable products
- increasing the capacity to innovate
- conservation of resources
- increasing organizational efficiency
- job creation
- clean production
- economic growth
- improving behavior in society.

Among the *barriers* that can be registered at the organizational level for the adoption of the circular economy are:

- organizational culture
- reduces organizational resources
- the complexity of the supply chain
- lack of identification of some benefits by the interested parties

- lack of knowledge in the field of sustainability and circular economy
- product design
- data uncertainty
- lack of ability to innovate.

Barriers can be overcome with the ability of stakeholders to identify organizational benefits. Stakeholders are sensitive to financial benefits. The circular economy also targets the *sustainable development goals* (SDG). Table 1.2 shows a correlation of the SDGs with the CE principles. The CE can contribute to the achievement of several sustainable development goals but aims at **SDG 12** (sustainable consumption and production).

1.4 The Importance of ICT for the CE

Globally, more than 4.9 billion people used the Internet in 2021. This percentage is continuously increasing. Therefore, information and communication technologies prove to be important for society and the industrial environment.

In Romania, from the perspective of the research-development activity, the enterprises from some fields of activity have more projects for the research-development activity, and others much less. The situation is presented in Annex 1, Table 1.4. From the perspective of the emissions that are generated, the situation presented in Annex 2, Table 1.5 can be noticed. These values represent values estimated by Eurostat and are ex-primate in thousand tons. At the level of the European Union, the number of emissions registered a decrease in the analyzed period, 2015–2020. At the level of Romania, there is a small decrease in certain years, the quantity is still over 90,000 tons/year. These situations underline the importance of adopting the principles of the circular economy [3, 7, 19].

The imports of waste for recovery—recycling (thousand tons) is presented in Annex, Table 1.6. At the level of the European Union the situation is fluctuating for the analyzed period 2015–2020. At the level of Romania, the important quantity is increasing compared to 2015. ICT is important is the CE. The following are some ICT guidelines for improving organizations' involvement in the CE. Below is an analysis of the action categories and their effect on the 11 R. We divide these action categories into 6 categories, as follows:

- C1—Green design—is essential to take the first steps in the CE at an organizational level.
- C2—Industry 4.0—outlines an extraordinary status for existing products on the market, including smart product, smart grids, and smart innovation.
- C3—Data analysis—internet of things, blockchain, digital platforms, artificial intelligence algorithms.
- C4—Software tools and simulation—are used for various actions of the circular economy. This category includes digital platform, PLM systems, smartphone application, and other software tools.

Table 1.2 Sustainable development goals and circular economy

Objective	Acronym	Circular economy
No poverty	SDG 1	It contributes to the eradication of poverty by reducing the number of resources used in the operational flow
Food	SDG 2	It contributes to the efficiency of resource consumption and thus contributes to achieving the goal of zero hunger
Health	SDG 3	The state of well-being is transposed especially from the perspective of the development of a clean environment, with a reduced amount of greenhouse gases. Thus, well-being and health are also covered by the CE
Education	SDG 4	CE contributes to a lesser extent to increasing the level of education. But, through its goals, the CE aims at job creation and cyclical involvement
Gender equality	SDG 5	The CE does not specifically refer to this SDG5
Water	SDG 6	Clean water and sanitation are also addressed by the CE
Energy	SDG 7	Energy efficiency is one of the CE's goals
Decent work and economic growth	SDG 8	Economic growth is targeted by the CE by streamlining the consumption of resources, innovation and creative jobs
Infrastructure	SDG 9	Industry, Innovation, and Infrastructure are CE targets
Reduces inequalities	SDG 10	It is less targeted by the CE, but indirectly touches on this aspect as well
Cities	SDG 11	Contribute to the development of sustainable and smart cities by stimulating innovation and reducing greenhouse gas emissions
Responsible consumption and production	SDG 12	It directly contributes to the achievement of this objective, being a priority for the circular economy
Climate	SDG 13	CE supports reducing greenhouse gas emissions and improving air quality
Oceans	SDG 14	It is a goal less addressed by the CE. Indirect actions can be identified
Life on land	SDG 15	Biodiversity is indirectly addressed by the CE
Institution	SDG 16	Peace, justice and strong institutions are addressed directly and indirectly by the CE
The power of partnerships	SDG 17	The implementation of partnerships for the efficiency of the activity and the elimination of waste are among the activities of the CE

Table 1.3 Green design, industry 4.0, data analysis, software tools and simulation, computing technologies, data management and storage, and 11 Rs

Rs	C1	C2	C3	C4	C5	C6
Reimage	X		X			X
Reduce	X	X	X	X	X	X
Reuse		X		X	X	X
Redesign	X	X	X	X	X	X
Recycle	X	X		X	X	X
Rethink	X	X	X	X	X	X
Repair			X		X	X
Recondition			X	X	X	X
Remanufacture			X		X	X
Recover	X	X	X	X	X	X
Reject	X	X	X	X	X	X

- C5—Computing Technologies—care include cloud computing, edge computing, thin client computers.
- C6—Data management and storage—blockchain, security, information management systems, smart connection.

Table 1.3 shows the implications of the 6 categories for the 11 Rs. Depending on the complexity of the category, it can contribute to the achievement of the 11 Rs or less.

1.5 Business Architecture for CE

Businesses are increasingly interested in applying the directions of action of the circular economy. The following is an architectural model for the relationship between ICT and the circular economy [12, 22, 29, 30]. This model includes the main modules to be evaluated and included in an organizational platform for the circular economy.

The main modules to be involved in an enterprise model are, Fig. 1.3:

- Data providers—includes product management and other facilities and laws of the field of activity
- Core platform—information on organizational operations
- Rs functions—includes the 11 functions
- life cycle assessment (LCA) module—refers to the organizational capacity to improve activities throughout the life cycle
- End user module—includes organizational evaluation based on the principles of sustainable development and the organizational capacity to develop sustainably.

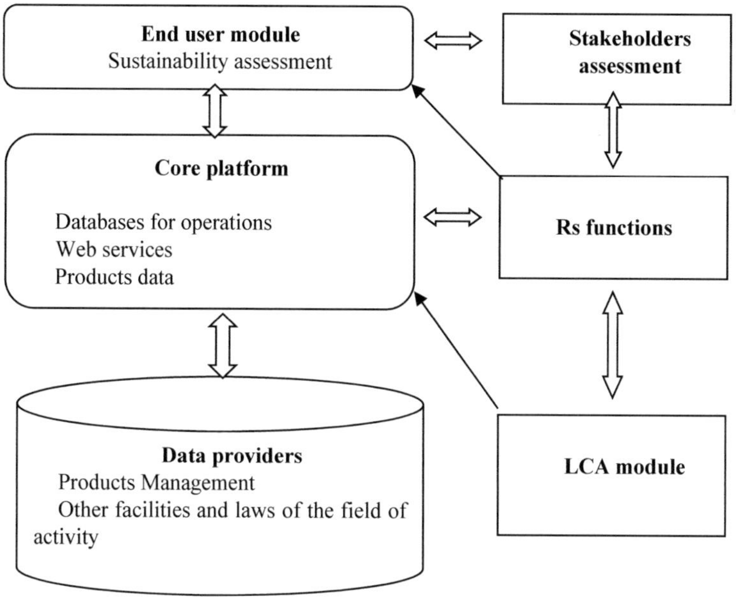

Fig. 1.3 Business architecture for circular economy

- Stakeholders—includes the inside and the outside. Their perception of ICT approaches to the circular economy will be assessed.

1.6 Conclusions

The chapter has provided a brief description of the concept of circular economy in the dynamics of the current developments. The implications of the 11 Rs in organizational activity have been shown to contribute to increasing the implications for sustainable development. The proposed model considers the needs of stakeholders and emphasizes the need to use technology for involvement in the circular economy. The principal aspects of the proposed model consider the 11 Rs of the circular economy and other components described in the previous part.

Annex 1

See Table 1.4.

Table 1.4 Projects for research and development, according to NABS

Dimension	Year					
	2015	2016	2017	2018	2019	2020
Exploration and exploitation of the land	25	55	60	119	121	65
Environment	479	501	659	366	286	431
Exploration and exploitation of space	73	33	62	61	63	45
Transport, telecommunications and other infrastructure	90	92	105	112	90	121
Energy	117	136	137	127	140	101
Industrial production and technology	736	783	739	549	477	489
Health	446	450	460	486	494	426
Agriculture	794	804	793	948	675	818
Education	300	312	421	263	290	264
Culture, recreational activities, religion and media	186	223	314	276	170	129
Political and social systems, structures and processes	119	142	110	140	144	88
General promotion of knowledge: research and development financed from the general university funds (GUF), for	1663	1710	1855	1776	1858	2282
Natural and exact sciences	348	295	320	187	1319	1444
Engineering and technological sciences	1094	1235	1359	1377	297	454
Medical and health sciences	43	70	36	12	44	58
Agricultural sciences	94	45	70	38	52	51
Social and economic sciences	59	56	49	131	130	212
Humanities sciences	25	9	21	31	16	63
General promotion of knowledge: research and development funded from sources other than general university funds, for	2763	2642	2669	3112	2597	2237
Natural and exact sciences	921	1130	869	1019	796	683
Engineering and technological sciences	1281	926	1076	1084	1028	958
Medical and health sciences	35	26	56	138	39	48
Agricultural sciences	31	52	43	67	22	31
Social and economic sciences	190	164	237	319	299	197
Humanities sciences	305	344	388	485	413	320

NABS = nomenclature for the analysis and comparison of scientific programmes and budgets
Source NIS [20], http://statistici.insse.ro:8077/tempo-online/#/pages/tables/insse-table

Annex 2

See Tables 1.5 and 1.6.

Table 1.5 Environment indicator—emission (thousand tonnes)

European Union—27 countries (from 2020)	2,656,120	2,641,104	2,624,124	2,572,758	2,569,780	2,567,866	2,598,378	2,524,663	2,512,639	2,498,354
European Union—28 countries (2013–2020)	2,920,657	2,910,420	2,896,199	2,833,921	2,824,189	2,814,157	2,844,514	2,767,492	2,750,884	:
Belgium	64,701	65,049	65,248	61,800	63,950	63,321	65,092	63,558	63,998	64,009
Bulgaria	48,004	43,890	40,004	35,277	38,240	35,945	38,039	36,953	34,161	33,141
Czechia	79,044	75,052	72,107	73,913	75,075	77,662	76,360	72,288	71,882	71,083
Denmark	51,068	48,719	48,309	46,293	46,238	47,407	48,680	47,875	48,270	48,779
Germany (until 1990 former territory of the FRG)	719,421	745,093	744,158	725,316	722,693	722,867	707,829	686,439	676,720	665,227
Estonia	15,993	15,879	16,910	16,458	15,228	16,518	17,713	17,126	17,293	17,810
Ireland	51,523	53,530	52,892	53,411	54,359	57,479	59,651	59,653	61,214	62,927
Greece	87,573	88,607	81,116	76,787	72,208	62,587	66,412	64,243	61,107	58,332
Spain	197,497	192,740	189,853	189,882	195,711	194,119	198,652	199,831	202,318	203,970
France	303,607	300,778	307,145	301,541	301,793	299,599	311,514	299,796	299,359	298,751
Croatia	14,584	13,704	13,831	14,061	13,810	14,613	14,612	14,643	13,465	13,379
Italy	247,838	237,982	226,447	224,102	224,083	222,230	225,909	224,914	225,117	225,376
Cyprus	4,093	3,826	3,545	3,610	3,577	3,568	3,699	3,690	3,710	3,743

(continued)

Table 1.5 (continued)

Latvia	10,279	10,777	10,713	11,104	11,057	11,014	11,330	11,697	11,738	11,909
Lithuania	13,767	14,271	14,235	13,893	13,928	13,877	14,252	14,285	14,383	14,497
Luxembourg	4,286	4,297	4,391	4,588	4,967	4,904	4,946	5,015	5,121	5,160
Hungary	42,910	41,091	41,891	42,389	42,063	42,434	42,257	41,336	41,073	40,825
Malta	1,564	1,653	1,548	1,751	1,648	1,613	1,688	1,784	1,792	1,828
Netherlands	124,433	122,235	123,628	122,091	126,173	123,994	123,426	117,529	116,388	113,942
Austria	69,472	69,673	76,586	74,233	72,961	71,962	72,482	67,006	66,498	64,882
Poland	236,666	236,426	233,903	224,713	218,595	228,202	237,572	222,344	221,751	222,540
Portugal	36,627	35,397	35,257	35,766	38,037	37,121	39,219	37,828	38,344	38,420
Romania	97,115	93,140	89,166	90,088	89,032	87,607	90,503	90,044	90,033	90,283
Slovenia	11,464	11,872	12,052	10,918	11,389	11,871	11,795	10,311	10,159	9,851
Slovakia	24,075	23,035	23,259	25,120	24,550	24,816	24,795	22,688	22,080	21,463
Finland	46,638	42,577	45,590	44,197	40,224	40,846	40,710	43,579	43,424	44,225
Sweden	51,879	49,812	50,338	49,458	48,189	49,690	49,241	48,209	47,897	47,824
United Kingdom	264,537	269,315	272,075	261,163	254,409	246,291	246,136	242,829	238,245	–

Source Eurostat [10] (https://ec.europa.eu/eurostat/databrowser/view/ENV_AC_SD__custom_1993163/default/table?lang=en)

Table 1.6 Imports of waste for recovery—recycling (thousand tonnes)

European Union—27 countries (from 2020)	12,904	12,990	13,424	13,742	13,324	12,465
European Union—28 countries (2013–2020)	8,510	9,108	9,511	9,916	10,033	–
Belgium	8,741	9,225	9,024	8,917	8,484	8,584
Bulgaria	356	397	437	484	580	476
Czechia	1,271	1,396	1,472	1,489	1,428	1,392
Denmark	598	665	739	786	850	858
Germany (until 1990 former territory of the FRG)	14,122	14,152	14,858	14,590	14,348	13,692
Estonia	127	172	196	234	271	253
Ireland	521	583	662	757	657	547
Greece	654	1,053	1,106	1,147	1,100	979
Spain	7,877	6,943	7,317	6,911	7,037	6,186
France	5,940	5,664	5,513	5,727	5,534	5,324
Croatia	286	369	354	448	483	657
Italy	6,644	6,555	7,324	7,848	7,466	7,095
Cyprus	1	3	9	11	10	6
Latvia	293	229	244	291	342	467
Lithuania	296	325	393	518	575	698
Luxembourg	2,695	3,248	2,800	2,977	3,028	2,783
Hungary	823	935	966	915	1,007	959
Malta	4	1	1	3	2	4
Netherlands	6,563	7,944	8,905	8,500	9,188	8,888
Austria	3,755	3,702	3,895	4,326	4,394	4,370
Poland	2,156	2,293	2,425	2,592	2,453	2,379
Portugal	2,275	2,158	2,438	2,478	2,129	2,215
Romania	309	544	553	640	661	558
Slovenia	1,221	1,318	1,291	1,309	1,233	1,177
Slovakia	472	551	742	811	540	713
Finland	323	343	348	394	421	441
Sweden	1,961	2,048	1,870	1,925	1,946	1,378
United Kingdom	0	0	0	0	0	–

Source Eurostat [10] (https://ec.europa.eu/eurostat/databrowser/view/ENV_AC_SD__custom_1993163/default/table?lang=en)

References

1. Abdul-Rashid, S.H., Sakundarini, N., Raja Ghazilla, R.A., Thurasamy, R.: The impact of sustainable manufacturing practices on sustainability performance: empirical evidence from Malaysia. Int. J. Oper. Prod. Manag. (2017). https://doi.org/10.1108/IJOPM-04-2015-0223
2. Abdullah, I.M., Dechun, H., Sarfraz, M., Ivascu, L., Riaz, A.: Effects of internal service quality on nurses' job satisfaction, commitment, and performance: mediating role of employee well-being. Nurs. Open J. **8**(2) (2020). https://doi.org/10.1002/nop2.665
3. Carney, M., Duran, P., van Essen, M., Shapiro, D.: Family firms, internationalization, and national competitiveness: does family firm prevalence matter? J. Family Bus. Strategy **8**(3), 123–136 (2017). https://doi.org/10.1016/j.jfbs.2017.06.001
4. Chen, Z., Chen, S., Liu, C., Nguyen, L.T., Hasan, A.: The effects of circular economy on economic growth: a quasi-natural experiment in China. J. Clean. Prod. **271**, 122558 (2020). https://doi.org/10.1016/j.jclepro.2020.122558
5. Cioca, L.I., Ivascu, L., Turi, A., Artene, A., Găman, G.A.: Sustainable development model for the automotive industry. Sustain. J. **11**(2), 6447 (2019). https://doi.org/10.3390/su11226447
6. Coste-Maniere, I., Croizet, K., Sette, E., Fanien, A., Guezguez, H., Lafforgue, H.: 6—Circular economy: a necessary (r)evolution. In: Muthu, A. (ed.) The Textile Institute Book Series, pp. 123–148. Woodhead Publishing (2019). https://doi.org/10.1016/B978-0-08-102630-4.00006-6
7. de Carvalho, R.A., da Hora, H., Fernandes, R.: A process for designing innovative mechatronic products. Int. J. Prod. Econ. **231**, 107887 (2021). https://doi.org/10.1016/j.ijpe.2020.107887
8. Duță, L., Filip, F.G.: Control and decision-making process in disassembling used electronic products. Stud. Inform. Control **17** (1) (2008)
9. Erdmann, L., Hilty, L., Goodman, J., Arnfal, P.: The future impact of ICTs on environmental sustainability. Institute for Prospective Technology Studies, Seville, Spain (2004)
10. Eurostat. https://ec.europa.eu/eurostat/databrowser/view/cpc_enclimwa/default/table?lang=en. Accessed 5 Dec 2021
11. Filip, F.G., Duță, L.: Decision support systems in reverse supply chain management. Procedia Econ. Finance **22** (2015). https://doi.org/10.1016/S2212-5671(15)00249-X
12. Geng, D., Lai, K., Zhu, Q.: Eco-innovation and its role for performance improvement among Chinese small and medium-sized manufacturing enterprises. Int. J. Prod. Econ. **231**, 107869 (2021). https://doi.org/10.1016/j.ijpe.2020.107869
13. Gupta, S.M., Ilgin, A.M.: Multiple Criteria Decision-Making Applications in Environmentally Conscious Manufacturing and Product Recovery. CRC Press, Boca Raton (2018)
14. Ilgin, M.A., Gupta, S.M.: Environmentally conscious manufacturing and product recovery (ECMPRO): a review of the state of the art. J. Environ. Manag. **91**, 563–591 (2010)
15. Istudor, I., Filip, F.G.: The innovator role of technologies in waste management towards the sustainable development. Procedia Econ. Finance **8** (2014). https://doi.org/10.1016/S2212-5671(14)00109-9
16. Ivascu, L., Sarfraz, M., Domil, A., Bogdan, O.: Assessment of country institutional factor on sustainable energy target achievement in European Union. Econ. Res. Ekonomska Istraživanja (2021). https://doi.org/10.1080/1331677X.2021.2002706
17. Kalar, B., Primc, K., Erker, R.S., Dominko, M., Ogorevc, M.: Circular economy practices in innovative and conservative stages of a firm's evolution. Resour. Conserv. Recycl. **164**, 105112 (2021). https://doi.org/10.1016/j.resconrec.2020.105112
18. Machado, C.G., Winroth, M.P., Ribeiro da Silva, E.H.D.: Sustainable manufacturing in Industry 4.0: an emerging research agenda. Int. J. Prod. Res. (2020). https://doi.org/10.1080/00207543.2019.1652777
19. Moretto, A., Macchion, L., Lion, A., Caniato, F., Danese, P., Vinelli, A.: Designing a roadmap towards a sustainable supply chain: a focus on the fashion industry. J. Clean. Prod. **193**, 169–184 (2018). https://doi.org/10.1016/j.jclepro.2018.04.273
20. National Institute of Statistics (NIS). http://statistici.insse.ro:8077/tempo-online/#/pages/tables/insse-table. Accessed 10 Jan 2022

21. Patwa, N., Sivarajah, U., Seetharaman, A., Sarkar, S., Maiti, K., Hingorani, K.: Towards a circular economy: an emerging economies context. J. Bus. Res. (2020). https://doi.org/10.1016/j.jbusres.2020.05.015
22. Richardson, B.C.: Sustainable transport: analysis frameworks. J. Transp. Geogr. (2005). https://doi.org/10.1016/j.jtrangeo.2004.11.005
23. Sarfraz, M., Ivascu, L., Cioca, L.I.: Environmental regulations and CO_2 mitigation for sustainability: panel data analysis (PMG, CCEMG) for BRICS nations. Sustain. J. **14**(1) (2022). https://doi.org/10.3390/su14010072
24. Schroeder, P., Anggraeni, K., Weber, U.: The relevance of circular economy practices to the sustainable development goals. J. Ind. Ecol. (2019). https://doi.org/10.1111/jiec.12732
25. Seok, H., Nof, S., Filip, F.G.: Sustainability decision support system based on collaborative control theory. Annu. Rev. Control **36**(1) (2012). https://doi.org/10.1016/j.arcontrol.2012.03.007
26. Shirvanimoghaddam, K., Motamed, B., Ramakrishna, S., Naebe, M.: Death by waste: fashion and textile circular economy case. Sci. Total Environ. **718**, 137317 (2020). https://doi.org/10.1016/j.scitotenv.2020.137317
27. Singh, C., Singh, D., Khamba, J.S.: Analyzing barriers of Green Lean practices in manufacturing industries by DEMATEL approach. J. Manuf. Technol. Manag. (2020). https://doi.org/10.1108/JMTM-02-2020-0053
28. Stahel, W.R.: The circular economy. Nature **531**, 435–438 (2016). https://doi.org/10.1038/531435a
29. Thu Truong, T.T., Kim, J.: Do corporate social responsibility activities reduce credit risk? Short and long-term perspectives. Sustainability (Switzerland) (2019). https://doi.org/10.3390/SU11246962
30. Tucker, P.: The impact of rest breaks upon accident risk, fatigue and performance: a review. Work Stress (2003). https://doi.org/10.1080/0267837031000155949
31. Zhou, X., Song, M., Cui, L.: Driving force for China's economic development under Industry 4.0 and circular economy: technological innovation or structural change? J. Clean. Prod. **271**, 122680 (2020). https://doi.org/10.1016/j.jclepro.2020.122680

Chapter 2
Mathematical Model of the Financial Sustainability of a Public University

Lucian-Ionel Cioca, Melinda Timea Fülöp, and Teodora Odett Breaz

Abstract Higher education worldwide has witnessed major changes across nations beginning with the last decade of the twentieth century. In the 1990s there appeared a lot of reforms which were deeply rooted in the new public administrative regulations. These had a major effect on the university management at the global level. The general objective of this scientific approach is to study the financial sustainability of a public university in terms of financial indicators. The first part of the research aimed at a scientific-theoretical basis that was based on qualitative research in order to thoroughly investigate the current state of knowledge. The qualitative part is completed by a quantitative study based of a mathematical model on the financial sustainability of Romanian universities. The results indicate all the premises to create a general model of sustainable university that can be applied, depending on the specifics of the university, and that contributes to a good development over an average period of such an institution.

Keywords Sustainability · Higher education · Financial position · Financial performance

L.-I. Cioca (✉)
Faculty of Engineering, Lucian Blaga University of Sibiu, Blv. Victoriei, 10, 550024 Sibiu, Romania
e-mail: lucian.cioca@ulbsibiu.ro

M. T. Fülöp
Faculty of Economics and Business Administration, Babeş-Bolyai University, 400591 Cluj-Napoca, Romania
e-mail: melinda.fulop@econ.ubbcluj.ro

T. O. Breaz
Faculty of Economic Sciences, "1 Decembrie 1918" University of Alba Iulia, Alba Iulia, Romania
e-mail: breaz.teodora@uab.ro

L. Ivascu et al. (eds.), *Intelligent Techniques for Efficient Use of Valuable Resources*, Intelligent Systems Reference Library 227,
https://doi.org/10.1007/978-3-031-09928-1_2

2.1 Introduction

Higher education worldwide has witnessed major changes across nations beginning with the last decade of the twentieth century. In the 1990s there appeared a lot of reforms which were deeply rooted in the new public administrative regulations. These had a major effect on the university management at the global level [1].

The European Commission has expressed its desire to modernize universities since 2006; at the same time the European Commission has established that modernization is considered to have—if universities want to contribute to the EU's goal of becoming a global and knowledge-based economic area—a fundamental importance. European universities have enormous potential, which unfortunately remains largely untapped due to rigid structures and various disabilities. As they are public universities, they cannot decide on the distribution of the budget, which can be distributed according to the needs of each university. Unleashing the vast reservoir of knowledge, talent, and energy requires immediate, profound, and coordinated changes: from the way systems are governed and managed to the way universities are run [2].

University funding has declined in recent years, despite additional burdens and an increasing number of students. This decrease in Romania may be due to the economic situation specific to an emerging country or even due to the fact that the situations for each year have not been reanalyzed and reevaluated. The structural change in university funding in recent years has also led universities to fund the tasks they have to perform on a permanent basis, with limited funding for the program. The financial situation of universities is aggravated, among other things, by the structural underfunding of the university building and the need to generate funding from third parties. The capacity of universities to develop strategies for the targeted and sustainable improvement of the the quality of courses offered and to provide long-term perspectives to young university students is severely limited as a result of this development. This also applies to the fulfillment of important research tasks that serve imperative interests. Part of the sustainability of the university agreement is that a course of study should not only be started, but also completed and is the basis for a good start in one's career.

If the aim of universities' policy is to guarantee successful studies and the development of young university students, then these objectives can only be achieved with the help of a reliable and sustainable funding structure. Furthermore, universities are important actors in using their expertise to help solve important national tasks in the long run while maintaining international competitiveness.

In modern societies, the state invariably plays a key role in financing education. However, states play this very different role in different areas of education. From the perspective of financial science, there are three questions regarding the commitment of the state. First, what are the reasons for public participation in education funding? Second, what distribution of public and private funding is appropriate in the various individual stages of education, from pre-school to post-secondary education? And finally, thirdly, what changes in the composition of public and private funding are appropriate?

The answers to these questions naturally depend on the economic objectives pursued by education policy. Practically, as in other policy areas, the economic objectives of education policy can be divided into objectives of efficiency and equity. In addition to the objectives of efficiency and equity, the promotion of economic growth is occasionally mentioned as an independent objective of educational policy. This is especially true in the context of demographic change, as education is linked to the hope that it will generate an increase in productivity sufficient to compensate for the decline in employment.

Blaug identified a fundamental external effect, and he argues that a functional democracy is not possible without proper education [3]. Freeman and Polasky [4] provide an additional argument for the positive externalities of education [5]. Knowledge has the property of non-rivalry, because the transfer of knowledge to others does not reduce the level of knowledge of the sender. If, however, the transfer of knowledge can only be observed by the persons involved, i.e. if the transfer cannot be verified by third parties, the purchasers of knowledge will not want to spend more than the pure technical costs of the transfer of knowledge. Therefore, those who acquire new knowledge are not sufficiently compensated for by those to whom this new knowledge is shared. Indeed, Arnott and Rouse [6] and Robertson and Symons [7] provide evidence that positive externalities are particularly pronounced in the early stages of education. These positive externalities of education establish a role for the state in the financing education. The state can fulfill this role by providing free education or subsidizing privately funded education. However, the role of the state weakens during the individual life cycle. Heckman and Klenow [8] express particular doubts about the existence of positive externalities in higher education. They conclude that there is no empirical basis to argue that higher education produces positive externalities. It is true that the achievements of university graduates in science and practice benefit the society and, therefore, non-academics in various ways.

For higher education, the effects of funding constraints have been in the minds of researchers for many years. The work of Kane [9, 10], Ellwood and Kane [11], and Card [12] shows that liquidity constraints explain the relatively low participation of lower-income groups in higher education. Cameron and Heckman [13], as well as Carneiro and Heckman [14] qualify this result by placing it in the context of the life cycle. The authors show that lower income levels groups have a lower participation in higher education. However, this educational disparity arises from long-term rather than short-term liquidity constraints. Even preschoolers do not have enough resources to provide them with adequate access to education. Thus, we note that liquidity constraints lead not only lead to a problem of efficiency but also to a problem of equity. Educational opportunities are much better in higher income groups than in lower income groups, if the latter are limited in liquidity.

Lately, a lot of research work has been conducted on university autonomy. The focus was also laid on financial sustainability in public sector units. Accountability was also given more importance and was more taken into account [15–20]. These attempts have attracted developments in aspects of university autonomy. However, there are too few studies to cover Romania and the situation of Romanian universities, so we considered it useful to present a mathematical model for the sustainability of Romanian universities.

2.2 Literature Review

Our paper starts from the analysis of the typology and characteristics of Romanian universities, but also from the role that the Ministry of National Education and Scientific Research (MENCS), the National Council for Financing Higher Education (CNFIS), and the Romanian Agency for Quality Assurance in Higher Education (ARACIS) have in the university environment in Romania. The national bodies listed above use performance indicators for different purposes: prioritization of higher education institutions and curricula (MENCS), funding of public universities (CNFIS), or evaluation of the quality of higher education (ARACIS).

Thus, our attention was directed to the performance and financial sustainability of universities in order to fulfill the main missions of a university, among which we list the mission of teaching, research, and services addressed to the community, etc.

The financial sustainability of a public institution such as universities aims at the ability to manage financial capacity in the short and long term, while maintaining the level of quality of services provided [19]. Accordingly, financial sustainability in universities could be described as the capacity to comply with the annual budgets without facing any restrictions. Consequently, the institution's revenues must be above the operating expenses [21]. Within the higher education institutions, financial sustainability implies the fact that these institutions should yield higher revenue as compared to the provision of educational services and activities. In other words, the income which is generated by these universities should be above the average requirements to meet the salary needs of staff, employees, and the acquisition of different services.

The application and measurement of performance require a definition of the term performance first, and secondly determining the role that universities have. Like accounting information generated in the private sector, public sector accounting information also aims to meet the information needs of categories of users, needs related to issues such as financial status, management performance of public institutions, or costs of services provided [22–25].

From an administrative point of view, performance is a good management of resources, efficiency in the use of financial, non-financial, and other material resources, such as space, computers, or libraries. Furthermore, the creation of internal and international relations, through European projects, is an important indicators for the financial performance.

The university performance can be related by financial indicators, but when we talk about the public sector that mostly provides services to citizens, having performance means meeting the needs of users that can be demonstrated by non-financial indicators.

Thus, defining performance is a complicated process, but it can be explained by reaching indicators. The funding that universities receive from the state budget, in addition to the fact that it has decreased [26–29], depends to a greater extent on the performance of higher education institutions [30–33].

Regarding presentation and terminology differences, although the information reported is similar (apart from recognition and measurement differences), the GFS and IPSAS use different names (e.g., 'Statement of Government Operations' versus "Statement of Financial Performance'), and the definition and/or value of key elements (such as total assets, net worth, total revenue and surplus/deficit) may differ [34, 35]. The classification structures of the information (financial statements) for the two reporting frameworks also differ.

In this regard, the GSR reports aim to provide a conceptual framework for the analysis and evaluation of fiscal policies and, in particular, the performance of the public sector at the level of each country, this system being developed for economic impact assessment of activities within each national economy [36–38]. There is a fundamental difference between budgetary and financial reporting, in that budgets are future-oriented financial plans for allocating resources, while financial statements describe, in retrospect, transactions and events in terms of performance and financial position [39].

Public universities are interested in achieving performance for two reasons [30]. The first would be to attract students, which is an important step in gaining an advantage in an increasingly competitive environment, by offering and meeting their needs. Universities are currently facing increasing competition in terms of attracting students. The second reason concerns public funding. Depending on the number of students enrolled, but also on the performance obtained by the university, public funds are allocated [40]. To achieve performance, a university must first meet the condition of providing quality services [41–44].

The Romanian education system has been influenced by political changes, starting with the 1990s, with the fall of communism, and continuing with the accession to the European Union. The Bologna process has brought the most consistent changes and has as its main objective the reduction of the discrepancy between the curricula of higher education institutions and the requirements of the labor market. European countries wanted to achieve the Single European Area in the field of education, to build comparable, compatible, coherent, and competitive higher education systems in all European countries.

2.3 Research Methodology

The general objective of this scientific approach is to study the financial sustainability of a public university in terms of financial indicators. The first part of the research aimed at a scientific-theoretical basis that was based on qualitative research in order to thoroughly investigate the current state of knowledge. Within the literature, there are similar current approaches to knowledge in the field of public sector accounting

[39, 45]. We consider that reviewing the specialized literature is an essential feature that any scientific approach must have, because an effective review of the relevant literature creates a solid theoretical foundation, which facilitates the identification of areas where research predominates and areas with additional research [46, 47] and is the foundation on which new discoveries are built [48, 49]. The identification and analysis of previous studies has helped us to create the premises for generating new knowledge, through a more rigorous approach that seeks to reduce the subjectivity of the researcher [50]. The qualitative part is completed by a quantitative study based of a mathematical model on the financial sustainability of Romanian universities.

2.4 Practical Implications for the Performance of Higher Education Institutions

When the question of what the performance of a higher education institution is, we often think of doing this analysis in terms of indicators that measure and highlight data on the quality of products offered to society, respectively, skills, in especially in the training of students and knowledge in various forms: fundamental and applied scientific research, consulting, expertise, involvement of members of the academic community in the life of society.

Defining sustainability is complicated, as this is another facet of the debate. The concept of sustainability refers to a sustainable development and may be seen as a mixture of three core elements that interact one with the other: the economic component, the environmental component and the social component.

We consider incomes and expenses in the period 2008–2021 (Table 2.1) and we will make using regression models a mathematical model with the help of which we can analyze and forecast the financial evolution of a university.

The budget of revenues and expenditures approved for 2015 included the financial resources necessary to finance the expenditures regarding the good development of the institution's activity. Thus, revenues and expenditures were based on the main specific indicators, namely: the number of subsidized full-time education students, the number of full-time education and fee-paying ID students, the number of subsidized and fee-based master's and doctoral students, the number of effective positions occupied, the total salary fund, the existing material base as well as the apparatus and equipment necessary for the educational process.

The 2016 revenue and expenditure budget for 2016 included the financial resources necessary to finance expenses related to the smooth running of the institution's activity. Thus, revenues and expenditures were based on the main specific indicators, namely: the number of full-time (part-time) students financed from the state budget, the number of full-time and distance education students for a fee, the number of master students and subsidized and paid doctoral students, the number of positions actually occupied, the total salary fund, the existing material base, as well as the apparatus and equipment necessary for the educational process. The revenue

Table 2.1 Incomes

Year	Initial prevision	Final previsions
2008	81,531,000	80,322,134
2009	57,308,109	60,246,123
2010	77,102,105	58,478,015
2011	56,199,245	64,565,002
2012	57,539,345	62,545,001
2013	59,870,234	57,32,123
2014	73,712,772	71,072,199
2015	25,757,774	31,073,648
2016	22,951,969	32,891,056
2017	40,778,523	44,026,906
2018	36,946,402	44,631,595
2019	55,173,245	66,260,962
2020	93,182,134	79,891,000
2021	100,947,000	86,548,000

and expenditure budget initially approved was rectified during the year as a result of the influences that appeared in the development of the activity and according to the institutional and complementary contracts concluded with the Ministry of National Education.

The budget of revenues and expenditures initially approved was rectified during the year as a result of the influences that appeared in the development of the activity. Thus, the initial provisions were of 22,951,969 lei, initial expenses at fee, final provisions 32,891,056 lei, and final expenses in the amount of 38,191,056 lei.

The revenue and expenditure budget of the institution is constituted on each source of financing, with respect to the budgetary classification. The difference between income and expenditure is due to the approval by the chief authorizing officer of the balance of the balance of the amount of 5.300,000 lei, amount allocated for Erasmus funds that were collected in 2015, and expenditure on staff and goods and services. The execution of the general budget of revenues and expenditures, calculated based on accounting data, is presented as detailed below.

The University income and expenditure budget of the University approved by the Ministry of National Education for 2017 included the financial resources necessary to finance expenses related to the good development of the institution's activity. Thus, revenues and expenditures were based on the main specific indicators, namely: the number of full-time (part-time) students financed from the state budget, the number of full-time and distance education students for a fee, the number of master students and subsidized and paid doctoral students, the number of positions occupied, the total salary fund, the existing material base, as well as the apparatus and equipment necessary for the educational process. The revenue and expenditure budget initially approved was rectified during the year because of the influences that appeared in the development of the activity and according to the institutional and complementary contracts concluded with the Ministry of National Education.

The budget of revenues and expenditures initially approved was rectified during the year because of the influences that appeared in the development of the activity. Thus, the initial provisions were 40,778,523 lei, initial expenses in the amount of 41,878,523, final provisions 44,026,906 lei and final expenses in the amount of 49,980,576 lei.

The budget of revenues and expenditures approved for 2018 included the financial resources necessary to finance the expenditures regarding the good development of the institution's activity. Thus, revenues and expenditures were based on the main specific indicators, namely: the number of subsidized full-time education students, the number of full-time education and fee-paying ID students, the number of subsidized and fee-based master's and doctoral students, the number of effective positions occupied, the total salary fund, the existing material base, as well as the apparatus and equipment necessary for the educational process.

The initially approved income and expenditure budget was rectified during the year because of the influences in the activity and according to the institutional and complementary contracts concluded with M.E.N, and to cover the deficit, the approval of the expenditure from the previous years' balance was requested.

The budget of revenues and expenditures initially approved was rectified during the year because of the influences that appeared in the development of the activity. Thus, the initial provisions were 36,946,402 lei, initial expenses 40,511,678 lei, final provisions 44,631,595 lei, and final expenses in the amount of 49,796,871 lei.

The difference between the provisions between income and expenses is due to the approval by the main authorizing officer of the balance of the balance of the amount of 5,165,276 lei, amount allocated for Erasmus funds that were collected in 2017 in the amount of 2,950,000 lei, expenses related to salaries in the amount of 1,000,000 lei, material expenses in the amount of 900,000 lei and investment expenses in the amount of 315,276 lei. In 2018, the institution was included in the payments made in the current year.

The budget of revenues and expenditures approved for 2019 included the financial resources necessary to finance the expenditures regarding the good development of the institution's activity. Thus, revenues and expenditures were based on the main specific indicators, namely: the number of subsidized full-time education students, the number of full-time education and fee-paying ID students, the number of subsidized and fee-based master's and doctoral students, the number of effective positions occupied, the total salary fund, the existing material base as well as the apparatus and equipment necessary for the educational process.

The initially approved income and expenditure budget was rectified during the year because of the influences in the activity and according to the institutional and complementary contracts concluded with MEC, and to cover the deficit, the approval of the expenditure from the previous years' balance was requested.

The budget of revenues and expenditures initially approved was rectified during the year as a result of the influences that appeared in the development of the activity.

Thus, the initial provisions were of 55,173,245 lei, initial expenses as well, final provisions of 66,260,962 lei, and final expenses in the amount of 66,260,962 lei.

The difference between the provisions between revenues and expenses is due to the approval by the main authorizing officer of the balance of the balance of the amount of 3,200,000 lei, the expenses related to salaries in the amount of 1,500,000 lei, the material expenses in the amount of 500,000 lei, expenses related to Erasmus funds which were collected in the previous year, and expenses with investments in the amount of 700,000 lei. In 2019, our institution was included in the payments made in the current year.

2.5 The Mathematical Model Regarding the Sustainability of Higher Education

We consider the budget of revenues and expenditures initially approved (initial provisions) rectified during the year as a result of the influences appearing in the development of the activity. Synthetically, the values for 2015–2019 are summarized in the table above.

From a mathematical point of view, the model can be presented as follows.

1. Approximation in the square mean

Let

$$f(x) = f(x; p_1, p_2, \ldots, p_m)$$

the function we want to obtain from the data set (xi, yi), i = (1, n)-. If we consider that the measurement errors are found in y, the most common method of approximation is to approximate the square mean, i.e., the problem of minimizing the function.

$$S(p_1, p_2, \ldots, p_m) = \sum_{i=1}^{n} [y_i - f(x_i)]^2 \tag{2.1}$$

Thus, the optimal values of the parameters are provided by the solutions of the equations.

$$\frac{\partial S}{\partial p_k} = 0, k = \overline{1, m} \tag{2.2}$$

The terms $r_i = y_i - f$ (xii) in relation (2.1) are called residues and represent the distance between the original points (measured) and the approximate values of the function f at the points x_i.

Remark 2.1 Usually Eq. (2.2) are nonlinear in p_j and thus difficult to solve. If we choose the approximation function as a linear combination for $f_j(x)$:

$$f(x) = p_1 f_1(x) + p_2 f_2(x) + \cdots + p_m f_m(x)$$

then Eqs. (2.2) become linear. The most natural way to choose polynomials $f_j(x)$ is:

$$f_1(x) = 1,\ f_2(x) = x,\ \ldots,\ f_m(x) = x^{m-1}$$

Remark 2.2 The distance from the approximation curve is called the standard deviation and is defined using the relation.

$$\sigma = \sqrt{\frac{S}{n - m}} \tag{2.3}$$

Remark 2.3 If we consider n = m, then the problem of approximation presented above is reduced to the problem of interpolation. This comes from the fact that the standard deviation considered in (2.3) cannot be defined.

2. Linear approximation

Let us consider the right

$$f(x) = p_1 + p_2 x.$$

In this case, the function to be minimized is

$$S(p_1, p_2) = \sum_{i=1} (y_i - p_1 - p_2 x_i)^2.$$

Equations (2.2) in this case are:

$$\frac{\partial S}{\partial p_1} = \sum_{i=1}^{n} -2(y_i - p_1 - p_2 x_i) = 2\left(-\sum_{i=1}^{n} y_i + np_1 + p_2 \sum_{i=1}^{n} x_i\right) = 0$$

$$\frac{\partial S}{\partial p_2} = \sum_{i=1}^{n} -2(y_i - p_1 - p_2 x_i)x_i = 2\left(-\sum_{i=1}^{n} x_i y_i + p_1 \sum_{i=1}^{n} x_i + p_2 \sum_{i=1}^{n} x_i^2\right) = 0$$

Dividing the equations by 2n, we obtain

$$p_1 + \overline{x} p_2 = \overline{y}$$

$$p_1 \overline{x} + \left(\frac{1}{n} \sum_{i=1}^{n} x_i^2\right) p_2 = \frac{1}{2} \sum_{i=1}^{n} x_i y_i$$

where

$\overline{x} = \frac{\sum_{i=1}^{n} x_i}{n}$ și $\overline{y} = \frac{\sum_{i=1}^{n} y_i}{n}$ represent the arithmetic means for x_i and y_i, respectively. Thus, we obtain the parameters.

$$p_1 = \frac{\overline{y}\sum_{i=1}^{n} x_1^2 - \overline{x}\sum_{i=1}^{n} x_i y_i}{\sum_{i=1}^{n} x_i^2 - n\overline{x}^2} \tag{2.4}$$

$$p_2 = \frac{\sum_{i=1}^{n} x_i y_i - n\overline{x}\overline{y}}{\sum_{i=1}^{n} x_i^2 - n\overline{x}^2} \tag{2.5}$$

Remark 2.4 To determine the parameters p_1 and p_2 we can also use the simplified form of relations (2.4) and (2.5)

$$p_2 = \frac{\sum_{i=1}^{n} y_i(x_i - \overline{x})}{\sum_{i=1}^{n} x_i(x_i - \overline{x})}$$

$$p_1 = \overline{y} - \overline{x}p_2$$

2.1 Multilinear Approximation

Let us consider the linear shape.

$$f(x) = p_1 f_1(x) + p_2 f_2(x) + \cdots + p_m f_m(x) = \sum_{j=1}^{m} p_j f_j(x)$$

where $f_j(x)$ is called basic functions and are predefined (see Remark 2.1). Substituting the relation (2.1), we obtain

$$S = \sum_{i=1}^{n}\left[y_j - \sum_{j=1}^{m} p_j f_j(x_i)\right]^2$$

Respectively, (2.2) becomes

$$\frac{\partial S}{\partial p_k} = -2\left\{\sum_{i=1}^{n}\left[y_i - \sum_{j=1}^{m} p_j f_j(x_i)\right] f_k(x_i)\right\} = 0, k = \overline{1, m}$$

We divide by −2 and obtain

$$\sum_{j=1}^{m}\left[\sum_{i=1}^{n} f_j(x_i) f_k(x_i)\right] p_j = \sum_{i=1}^{n} f_k(x_i) y_i, k = \overline{1, m}$$

This relation can be written in matrix form.

$$\mathrm{Ap} = \mathrm{b} \tag{2.6}$$

where

$$A_{kj} = \sum_{i=1}^{n} f_j(x_i) f_k(x_i) b_k = \sum_{i=1}^{n} f_k(x_i) y_i \tag{2.7}$$

Equations (2.6) are called normal approximation equations.

(x)(y) calculations, are as follows: mean, respectively, line, Thus, obtain the following:

For a quick analysis of the results using the linear and multilinear mean, respectively, to determine the regression line, we will further use the cftool package from the MATLAB language.

Linear and multilinear analysis.

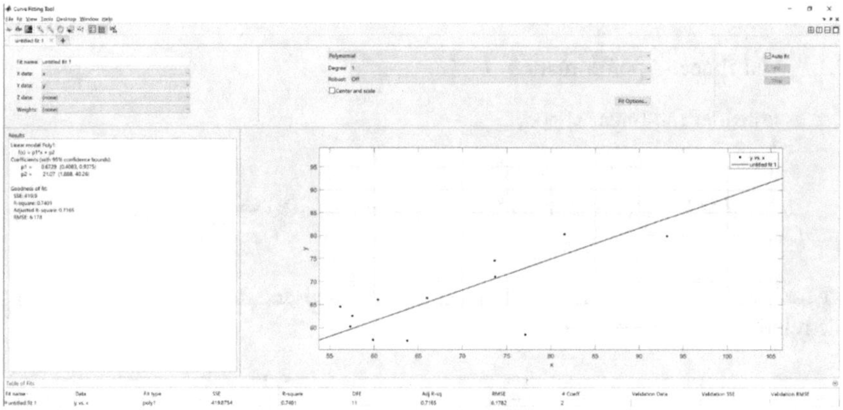

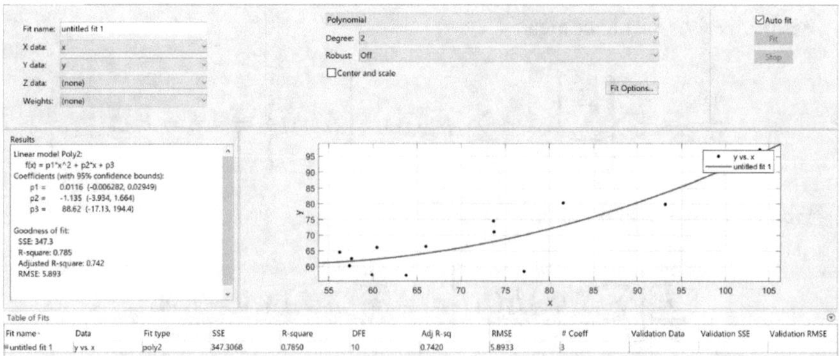

One can see that in both models R-square is quiet low, i.e. 0.7401 for the linear model and 0.785 for the multilinear model.

Prevision

Model validation was performed for the 2021 academic year. One can see that using the initial prevision in the linear model we obtain a final previous of 88.997 lei, which has a deviation of −2.448 million.

This is due to the low correlation of the linear model.

Comparative analysis using the sum-of-sine model.

One can easily see that using the sum of sine model with three terms, i.e.

$$\begin{aligned} f(x) = {} & a1 * \sin(b1 * x + c1) + a2 * \sin(b2 * x + c2) \\ & + a3 * \sin(b3 * x + c3) \end{aligned}$$

we obtain a better correlation for the model. In this case, the R-square is 0.9087.

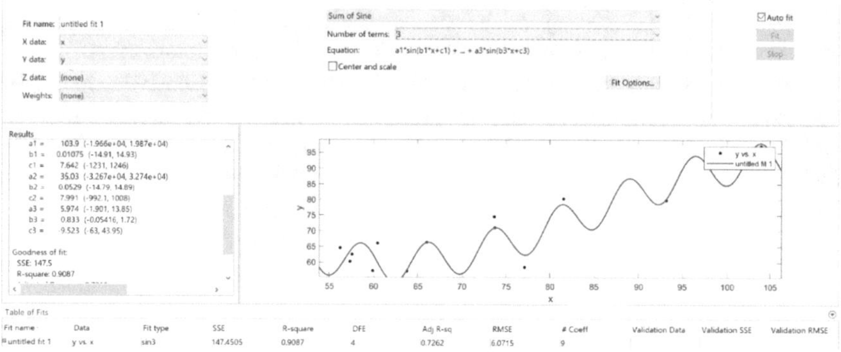

This also verify our foresight, in this case using the sum of sine model we obtain a final prevision of 86.578 million lei, which has a deviation of −0.029 million lei.

2.6 Conclusions

Based on the research, we notice an increase in funding and short-term and temporary programs of universities. On the other hand, it is necessary to anchor the measures permanently, constitutionally, and in the planning and management of universities. The state plays an important role in all phases of the educational life cycle. The role of the state is essentially based on the lack of rationality in the demand for education, the positive externalities of education, incomplete private credit markets for financing education, and the lack of equity in terms of educational opportunities. Investments in early education increase the productivity of further educational investments, which is why investments in early education are the most productive. Therefore, this effect is also called the self-productivity of education.

If one compares the educational policy, which is oriented towards efficiency and equity objectives, with the real educational policy, the deviations from the preschool sector and from the university become obvious. The reasoning so far would suggest more public funding in the preschool sector and more private funding in the higher education sector. It is true that there is often an objection to more private funding of higher education, i.e. tuition fees, that it has a selective social effect and is therefore not fair. In fact, this argument does not recognize that the social selection observed in universities is not triggered by tuition fees. Social selection is made much earlier in the life cycle. Several current studies suggest that the prerequisites for further educational success are already established at preschool age. Therefore, the state should start from there and offer more free educational offers for preschoolers. Higher education, on the other hand, can be funded more privately, without being linked to efficiency or equity issues.

This is supported by the fact that, while maintaining international competitiveness, universities contribute their skills to solving important national long-term and long-term tasks. In detail, z. Examples include digitization, climate research, the ongoing consolidation of other relevant research priorities, and internationalization.

There are all the prerequisites to create a general model of sustainable university that can be applied, depending on the specifics of the university, and that will contribute to a good development over an average period of such an institution. The implementation of questionnaires with the help of which to analyze the degree of student satisfaction in relation to the university is necessary.

References

1. Modugno, G., Di Carlo, F.: Financial sustainability of higher education institutions: a challenge for the accounting system. In: Financial Sustainability of Public Sector Entities, pp. 165–184. Palgrave Macmillan, Cham (2019)
2. European Commission. https://ec.europa.eu/commission/presscorner/detail/en/IP_06_592
3. Becker, G.S., Michael, G., Michael, R.T.: Economic Theory. Routledge (2017)
4. Freeman, S., Polasky, S.: Knowledge-based growth. J. Monetary Econ. **30**, 3–24 (1992)
5. Barkhordari, S., Fattahi, M., Azimi, N.A.: The impact of knowledge-based economy on growth performance: evidence from MENA countries. J. Knowl. Econ. **10**(3), 1168–1182 (2019)
6. Lorenz, G., Boda, Z., Salikutluk, Z., Jansen, M.: Social influence or selection? Peer effects on the development of adolescents' educational expectations in Germany. Br. J. Sociol. Educ. **41**(5), 643–669 (2020)
7. León, S.P., Augusto-Landa, J.M., García-Martínez, I.: Moderating factors in university students' self-evaluation for sustainability. Sustainability **13**, 4199 (2021)
8. Garrigos-Simon, F.J., Botella-Carrubi, M.D., Gonzalez-Cruz, T.F.: Social capital, human capital, and sustainability: a bibliometric and visualization analysis. Sustainability **10**(12), 4751 (2018)
9. Aina, C., Baici, E., Casalone, G., Pastore, F.: The determinants of university dropout: a review of the socio-economic literature. Socio-Econ. Plann. Sci. 101102 (2021)
10. Skinner, B.T.: Choosing college in the 2000s: an updated analysis using the conditional logistic choice model. Res. High. Educ. **60**(2), 153–183 (2019)

11. Iborg, D.H.: Examining the effects of ACT assessment of high school graduates on college enrollment and college readiness. Doctoral dissertation, Lindenwood University (2014)
12. Bailey, D.H., Duncan, G.J., Cunha, F., Foorman, B.R., Yeager, D.S.: Persistence and fade-out of educational-intervention effects: mechanisms and potential solutions. Psychol. Sci. Public Interest **21**(2), 55–97 (2020)
13. Biewen, M., Tapalaga, M.: Life-cycle educational choices in a system with early tracking and 'second chance' options. Econ. Educ. Rev. **56**, 80–94 (2017)
14. Decker, D.: Student perceptions of higher education and apprenticeship alignment. Educ. Sci. **9**(2), 86 (2019)
15. Ödemiş, S., Beytekin, O.F., Uslu, M.E.: An exploratory study on university autonomy: a comparison of Turkey and some European Union countries. Educ. Instruct. Stud. World **6**, 79–90 (2016)
16. Rodríguez Bolívar, M.P., Navarro Galera, A., Alcaide Muñoz, L., López Subirés, M.D.: Risk factors and drivers of financial sustainability in local government: an empirical study. Local Gov. Stud. **42**(1), 29–51 (2016)
17. Bisogno, M., Cuadrado-Ballesteros, B., García-Sánchez, I.M.: Financial sustainability in local governments: definition, measurement and determinants. In: Financial Sustainability in Public Administration, pp. 57–83. Palgrave Macmillan, Cham (2017)
18. Ferry, L., Murphy, P.: What about financial sustainability of local government!—a critical review of accountability, transparency, and public assurance arrangements in England during Austerity. Int. J. Public Adm. **41**(8), 619–629 (2018)
19. Caruana, J., Brusca, I., Caperchione, E., Cohen, S., Rossi, F.M. (eds.): Financial Sustainability of Public Sector Entities: The Relevance of Accounting Frameworks. Springer (2019)
20. Lee, Y.H., Kim, K.S., Lee, K.H.: The effect of tuition fee constraints on financial management: evidence from Korean private universities. Sustainability **12**(12), 5066 (2020)
21. Xu, W., Fu, H., Liu, H.: Evaluating the sustainability of microfinance institutions considering macro-environmental factors: a cross-country study. Sustainability **11**(21), 5947 (2019)
22. Tallaki, M.,Bracci, E.: Risk perception, accounting, and resilience in public sector organizations: a case study analysis. J. Risk Financ. Manag. **14**(1), 4 (2021)
23. Haustein, E., Lorson, P.C., Caperchione, E., Brusca, I.: The quest for users' needs in public sector budgeting and reporting. J. Public Budg. Account. Financ. Manag. (2019)
24. Dabbicco, G.: The potential role of public sector accounting frameworks towards financial sustainability reporting. In: Financial Sustainability of Public Sector Entities, pp. 19–40. Palgrave Macmillan, Cham (2019)
25. Mihut, M.I., Crisan, A.R.: Harmonisation, a road to the public sector accounting modernisation? Int. J. Econ. Account. **9**(3), 221–237 (2020)
26. Almohtaseb, A.A., Almahameed, M.A.Y., Shaheen, H.A.K., Al Khattab, M.H.J.: A roadmap for developing, implementing and evaluating performance management systems in Jordan public universities. J. Appl. Res. Higher Educ. (2019)
27. Adams, C.A.: Sustainability reporting and performance management in universities: challenges and benefits. Sustain. Account. Manag. Policy J. (2013)
28. Nisio, A., Carolis, R.D., Losurdo, S.: Introducing performance management in universities: the case of a university in Southern Italy. Int. J. Manag. Educ. **12**(2), 132–153 (2018)
29. Mouritzen, P.E.,Opstrup, N.: Performance Management at Universities. Springer International Publishing (2020)
30. Murias, P., de Miguel, J.C., Rodríguez, D.: A composite indicator for university quality assessment: the case of Spanish higher education system. Soc. Indic. Res. **89**(1), 129–146 (2008)
31. Emrouznejad, A., Yang, G.L.: A survey and analysis of the first 40 years of scholarly literature in DEA: 1978–2016. Socioecon. Plann. Sci. **61**, 4–8 (2018)
32. El Gibari, S., Gómez, T., Ruiz, F.: Evaluating university performance using reference point based composite indicators. J. Informet. **12**(4), 1235–1250 (2018)
33. Stumbriene, D., Camanho, A.S., Jakaitiene, A.: The performance of education systems in the light of Europe 2020 strategy. Ann. Oper. Res. **288**(2), 577–608 (2020)

34. André, P.: The role and current status of IFRS in the completion of national accounting rules–evidence from European countries. Account. Eur. **14**(1–2), 1–12 (2017)
35. Jesus, M.A., Jorge, S.: Accounting basis adjustments and deficit reliability: evidence from Southern European countries. Revista de Contabilidad **19**(1), 77–88 (2016)
36. Barton, A.: Why governments should use the government finance statistics accounting system. Abacus **47**(4), 411–445 (2011)
37. Crişan, A.R., Fülöp, M.T.: An analysis of the international proposals for harmonization accounts statement and government finance statistics. Account. Manag. Inf. Syst. **13**(4), 800–819 (2014)
38. Abbas, S.A., Pienkowski, A., Rogoff, K. (eds.): Sovereign Debt: A Guide for Economists and Practitioners. Oxford University Press (2019)
39. van Helden, J., Reichard, C.: Cash or accruals for budgeting? Why some governments in Europe changed their budgeting mode and others not. OECD J. Budg. **18**(1), 91–113 (2018)
40. McLendon, M.K., Hearn, J.C., Deaton, R.: Called to account: analyzing the origins and spread of state performance-accountability policies for higher education. Educ. Eval. Policy Anal. **28**(1), 1–24 (2006)
41. Dzimińska, M., Fijałkowska, J., Sułkowski, Ł: Trust-based quality culture conceptual model for higher education institutions. Sustainability **10**(8), 2599 (2018)
42. Varouchas, E., Sicilia, M.Á., Sánchez-Alonso, S.: Academics' perceptions on quality in higher education shaping key performance indicators. Sustainability **10**(12), 4752 (2018)
43. Hadullo, K., Oboko, R., Omwenga, E.: A model for evaluating e-learning systems quality in higher education in developing countries. Int. J. Educ. Dev. Using ICT **13**(2) (2017)
44. Olmos-Gómez, M.D.C., Luque Suarez, M., Ferrara, C., Olmedo-Moreno, E.M.: Quality of higher education through the pursuit of satisfaction with a focus on sustainability. Sustainability **12**(6), 2366 (2020)
45. Rajib, M.S.U., Hoque, M.: A literature review on public sector accounting research. Jahangirnagar J. Bus. Stud. **5**(1), 39–52 (2016)
46. Mohajan, H.K.: Qualitative research methodology in social sciences and related subjects. J. Econ. Dev. Environ. People **7**(1), 23–48 (2018)
47. Rajib, M.S.U., Hoque, M.: Application of ICT in public sector accounting of Bangladesh. Dhaka Univ. J. Bus. Stud. Spec. Int. Ed. **1**(1), 1–16 (2017)
48. Massaro, M., Dumay, J., Guthrie, J.: On the shoulders of giants: undertaking a structured literature review in accounting. Account. Audit. Account. J. **29**(5), 767–801 (2016)
49. Di Vaio, A., Palladino, R., Hassan, R., Escobar, O.: Artificial intelligence and business models in the sustainable development goals perspective: a systematic literature review. J. Bus. Res. **121**, 283–314 (2020)
50. Dumay, J., Bernardi, C., Guthrie, J., Demartini, P.: Integrated reporting: a structured literature review. In: Accounting Forum, vol. 40, no. 3, pp. 166–185. Elsevier (2016)

Chapter 3
Using Modern Information and Communication Technologies for Intelligent Capitalization of Cultural Resources

Cristian Ciurea and Florin Gheorghe Filip

Abstract The chapter presents the concepts and technology of virtual exhibitions, with a particular emphasis on the changes brought on by the current COVID-19 pandemic Modern information and communication technologies are used to promote a new paradigm for facilitating the access to cultural heritage collections, by implementing virtual exhibitions, in the form of web applications, accessible online, as well as native mobile applications for mobile devices. A practical example is presented. The *resource-based view* (RBV) concept is used to determine the strategic resources that cultural institutions can exploit to achieve sustainable competitive advantage.

Keywords GLAM · Smart city · Modern ICT · Cultural heritage · Resource-based view

3.1 Introduction

Modern information and communication technologies have enabled the digital transformation of cultural heritage into a strategic economic resource for European countries. The influences of multiculturalism, globalization, and technological revolution on cultural institutions, such as art *galleries, libraries, archives, museums* (GLAM) have led to major changes in the way they preserve, promote, and capitalize on cultural goods. The digitization of the cultural goods preserved in valuable heritage collections and the evolution of *information and communication technologies* (ICT) have determined the progress of virtual exhibitions as a tool for promoting and capitalizing on cultural heritage objects. New business models have been developed to

C. Ciurea
Department of Economic Informatics and Cybernetics, Bucharest University of Economic Studies, Bucharest, Romania
e-mail: cristian.ciurea@ie.ase.ro

F. G. Filip (✉)
The Romanian Academy, Bucharest, Romania
e-mail: ffilip@acad.ro

L. Ivascu et al. (eds.), *Intelligent Techniques for Efficient Use of Valuable Resources*, Intelligent Systems Reference Library 227,
https://doi.org/10.1007/978-3-031-09928-1_3

capitalize on cultural heritage in libraries and museums, in the context of globalization and the technological revolution, and the premises have been created for the development of a new category of entrepreneurs and businesses. Studies have been carried out on the increasing the number of visitors to cultural institutions using promotional tools, such as virtual exhibitions and tours, and dedicated mobile applications. The digitization process of cultural goods has got traction under the current circumstances created the pandemic when remote access has become preferred.

The chapter presents a methodology for assessing the economic potential generated by heritage collections of cultural institutions, as well as the development of economic models to capitalize heritage collections, in order to increase the public's interest in culture and, at the same time, to preserve the national identity awareness.

3.2 The Evolution of Modern Information and Communication Technologies and Their Impact to Cultural Institutions

According to [1], one of the most difficult aspects of creating a resilient community is that success is defined by what does not happen, a phenomenon that is difficult to quantify to some extent.

The COVID-19 pandemic has proven the importance of all companies being prepared to survive substantial and unanticipated change. Organizations that can successfully absorb and adapt to the challenges caused by the COVID-19 pandemic are *resilient*. Many people are also making judgments at a rate that would have been inconceivable prior to COVID-19 pandemic. Across all the organizations, five characteristics were identified in order to improve their agility [2]:

- Establish a common goal and open lines of communication;
- Create frameworks that allow for quick decision-making, including resource reallocation to meet changed priorities;
- Build networks of local teams with well-defined roles and responsibilities;
- Create a culture that empowers employees and enables them to pursue their entrepreneurial dreams;
- Make sure that individuals have access to the technologies they require.

Every aspect of human life has been altered by the fourth technology revolution. The cultural field has also had to adapt itself to the target audience's needs and behavioral-cultural profile. *Art galleries, libraries, archives, and museums* (GLAM), which are established to secure the infrastructure for the collection and preservation of a nation's values, have been adapting to the new *Information and Communication Technologies* (ICT) such as web and mobile technologies, *Cloud Computing, Internet of Things, Big Data,* and so on [3].

Cultural heritage has now become a strategic economic resource for European countries thanks to new information communication and technology. The potential market of cultural and research content contained in cultural institutions in Europe was estimated as 27 billion euros at the end of 2017 [4]. Modern ICT not only ensure better knowledge and protection of cultural assets, but also create the premise more direct visitors to cultural institutions. The aim of remote access is not to drive away people from real cultural institutions, but rather to stimulate them to visit them. Modern technologies, such as augmented and virtual reality, Big Data, Artificial Intelligence, the Internet of Things, Cloud, mobile technologies, have been adopted by all cities around the world in an attempt to create intelligent digital reality [5].

Human behavior has evolved as more people live in *smart cities* (SC) and benefit from modern ICT. Because of the technology has ensured comfort, a number of people are less inclined to interact personally and prefer to communicate via electronic means. Technology, on the other hand, can be one possible answer to overcoming loneliness and bringing people together through specific applications. The term "smart cities" has been superseded with "digital cities" in recent years [6, 7]. The improved access to cultural heritage is one of most relevant components of a Smart City. As one of the most dynamic sectors, cultural heritage contributes around 2.6% to EU GDP and employment (over 8 million jobs) [4]. The concept of "The Economy of Digital Culture" has come into common use. The term "e-Culture" is frequently used in Europe in the context of the information society, in addition to e-government, e-health and other e-domains [8]. The economy of culture generates a competitive advantage for new activities that exploit the value of cultural goods, through the use of modern ICT. Augmented reality, virtual reality, mobile applications and cloud platforms provide an effective opportunity to promote, preserve and protect heritage around the world.

Some countries have made efforts to maintain their cultural identity by promoting the most appealing cultural artefacts that symbolize their culture. The digitization of collections is critical not only for the long-term preservation of rare manuscripts, papers, seals, and paintings, but also for the creation of digital content for online exhibitions, virtual museums, and other multimedia representations of cultural collections, to increase their promotion and capitalization. As a link between globalization and culture, a new notion called "global citizenship" has evolved [9]. In this context, recovering and distinguishing a nation's cultural identity has become a more difficult undertaking. The amount of cultural collections that have been digitized has been used as a measure of cultural innovation. Only ten percent of European institutions' entire cultural heritage collections have been digitized. In the coming years, the remaining 90% should be digitized [10].

Cultural institutions grouped under the name GLAM are not only stores of knowledge through the preservation of heritage collections, but also creators of new knowledge through the use of current information and communication technologies (virtual exhibitions, virtual and 3D tours, etc.). The following are the goals of the virtual exhibitions:

- the primary goal is to promote and capitalize on heritage collections in cultural institutions;

- secondary goals include attracting new visitors, boosting public interest in specific collections, and increasing cultural institutions' revenue.

The EU Council confirmed in May 2014 that Europe's cultural heritage is a strategic asset. It is now viewed as a collection of tangible, intangible, and digital resources inherited from the past in all forms and aspects, including monuments, sites, skills, practices, knowledge, and expressions of human creativity, as well as related collections managed by public and private bodies, such as museums, libraries, and archives. Culture may be viewed the post-COVID19 world's "social cement" [4]. The cultural and creative sectors may prove to be the most important ally in Europe's economic revival. It is hoped that modernizing cultural institutions will improve their public image, visibility, number of visitors, and, most importantly, profits. Many cultural institutions require assistance in adapting to the present technological revolution and utilizing the benefits of modern ICT. One in every eight museums globally may never reopen as a result of the COVID19 crisis.

3.3 Virtual Exhibitions as a Tool for Promoting and Capitalizing on Heritage Objects

A virtual exhibition efficiently transposes the essence of cultural objects presented in physical format into the digital world, making it accessible to visitors from anywhere and at any time. Virtual exhibitions not only provide a better way to publicize cultural institutions' unique collections of books, old and rare manuscripts, paintings, sculptures, and other cultural objects, but they also provide some benefits to cultural institutions, such as increasing visitor numbers, improving visibility and public image, and increasing revenue from visitor entrance fees [10].

Depending on how the virtual exhibition is implemented, curators can decide which tool/technology could be most effectively used for digitization. Virtual exhibitions can be developed in the form of web applications, available online, with a responsive interface that adapts to each mobile device. In order to benefit from the new technologies such as sensors, and new *Near-Field Communication* [11] available for a mobile device, a virtual exhibition as a native mobile application should be created on Android and iOS operating systems. In this case, visitors can combine physical and virtual experiences, for example, when scanning QR (Quick Response) codes assigned to cultural objects in a physical exhibition (Fig. 3.1).

Multiculturalism, globalization, and the technological revolution have all had a significant impact on how GLAM selects, preserves, promotes, and capitalizes on cultural commodities. A country's economic development is a key factor that both stimulates and is stimulated by cultural development.

Studies have been conducted to see if promotional IT tools, such as virtual exhibitions and dedicated mobile applications, might help increase the number of visitors to cultural organizations. Depending on the level of economic growth, the culture is both steady and dynamic. As a country's wealth grows, it will see profound cultural

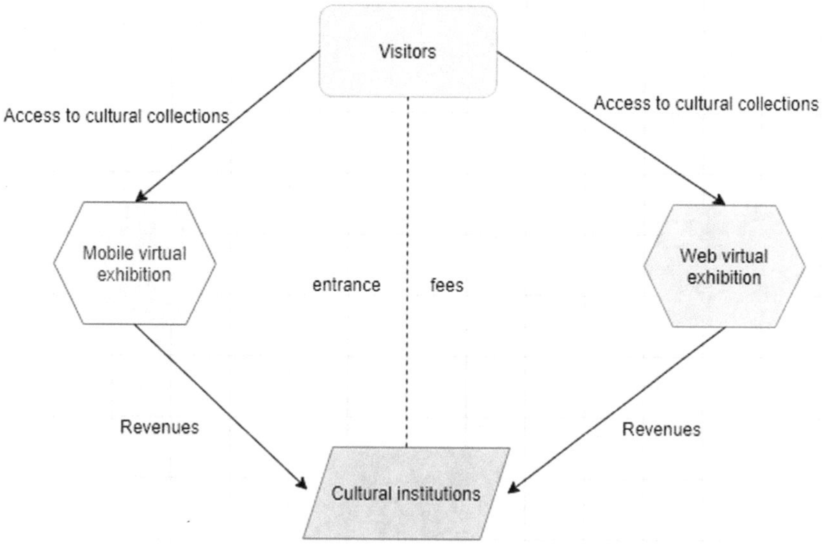

Fig. 3.1 Facilitating the access to cultural heritage collections

shifts in many aspects of life. In order to revitalize the cultural industry, not only ICT tools and techniques are used, but also some marketing strategies are necessary [12]. Virtual exhibitions include components from a variety of disciplines, including information technology and marketing. Implementing a virtual exhibition entails not just displaying images of cultural objects in an image gallery, but also making the complete product more appealing to visitors.

The virtual exhibition "Seals—the history treasure", which presents the collection of the Romanian Academy Library seals, was created as a web application using the MOVIO platform (https://github.com/GruppoMeta/Movio). The exhibition is available online at http://movio.biblacad.ro/SEALS/ and can be accessed both on computers and mobile devices, because it has a responsive interface that adapts depending on the screen size.

Figure 3.2 shows the interface of the virtual exhibition, which includes menu options to navigate through the entire collection of seals. It starts with a description ("About the seals"), and after that it presents categories such as "Pendant seals", "Attached seals" and "Impress seals".

Aside from royal seals, nobility (boyars) seals have been described, with signs influenced by medieval bestiary (lions, eagles, griffins, snakes, and so on), but the most fall into the geometric category. The bulk of seals were circular in shape in the past, but ovals, triangles, shield-shapes, and other patterns have also been seen. A graphic symbol (often, but not usually, integrating heraldic elements) was surrounded by text (the legend) that ran around the edge of the artwork. It's worth noting that

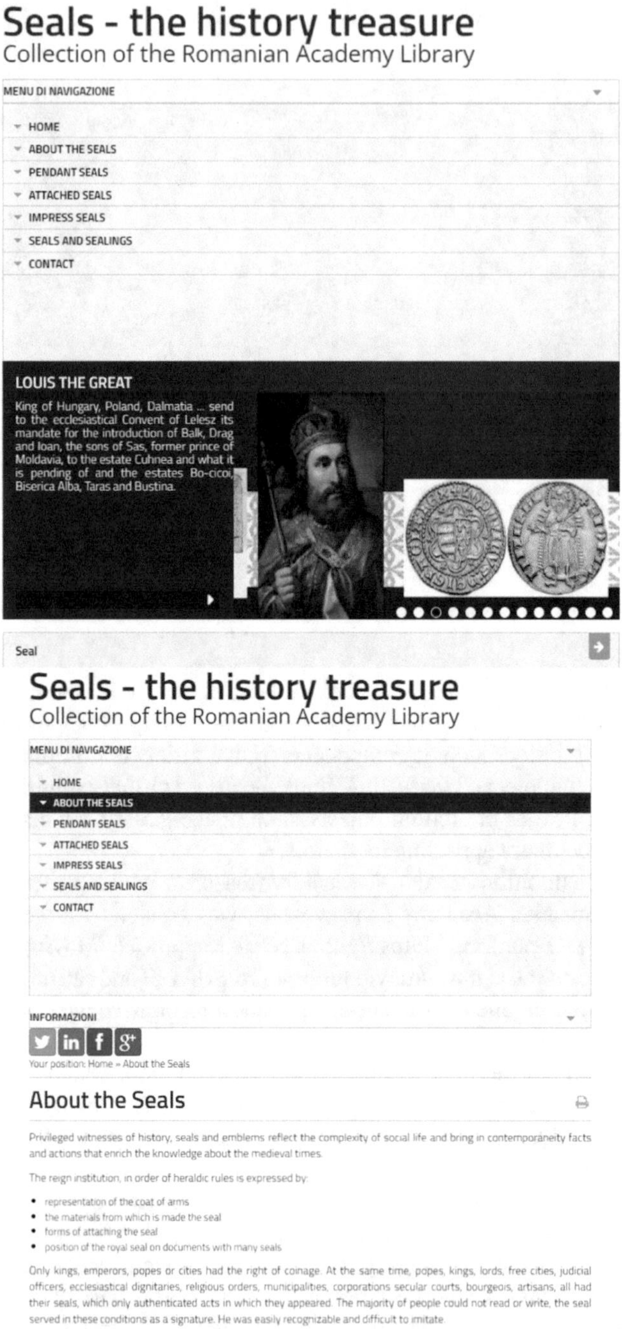

Fig. 3.2 Virtual exhibition "Seals—the history treasure": the interface

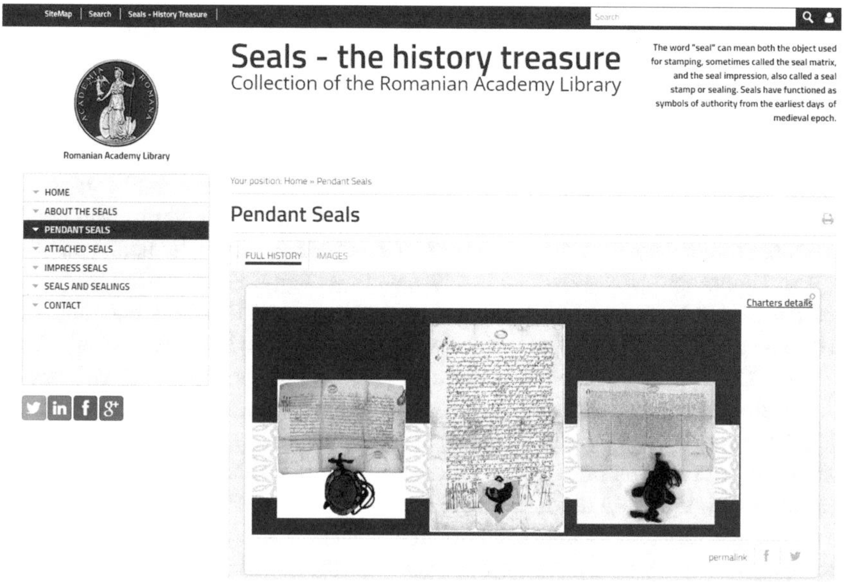

Fig. 3.3 Virtual exhibition "Seals—the history treasure": the main menu

the majority of the seals we're talking about have heraldic significance, even if the field mark is sometimes visible in the circular stamp.

The menu of the virtual exhibition seen from a computer screen is as in Fig. 3.3. It is placed in the left side of the screen. All additional pages are opened in the middle of the screen. On the top, the title and subtitle are always displayed. Under the menu, there are default buttons to share content via social networks.

The navigation through the exhibition will be recorded using Google Analytics tools, so that to analyze the user behavior regarding the most attractive collections. Also, we will track information about the countries from which the exhibition was accessed and the types of devices that were used (based on the operating system). These data will help the curators and developers to deliver personalized content in the future, based on the users' interests.

Several concepts such as *smart cultural heritage* and *cultural sustainability* could be defined. *Smart cultural heritage* is the combination of smart cities and visualization technologies used to access and promote cultural heritage, as cultural heritage has the potential to play a significant role in the development of smart cities. *Cultural sustainability* is the ability of cultural institutions to benefit economically, socially, and environmentally.

Nowadays it is necessary to publicize Europe's rich cultural heritage, particularly among young people. Storytelling is apparently the most effective way to add value to cultural collections, especially when they are accessed and augmented with modern information and communication, because it creates narrative experiences for visitors in the digital age. Virtual exhibitions are one of the many ICT deployments that are compatible with *Digital Humanism* key concepts [13].

When it comes to Digital Humanism, libraries, galleries, archives and museums (GLAM) may become indispensable as both a model and a testing ground. Inquiring about their own roles as gatekeepers and curators, the digital transition provides them with the opportunity to open up—through collaborative initiatives and comprehensive collections. However, it is also a question of sustaining the library and archive as a space of contact and personal conversation in the human and humanist tradition. A "third approach" of digitization is advocated by Digital Humanism. This implies that there must be a viable alternative to Tech Startups and Beijing, one that is not driven by greed or authoritarianism, but rather by a desire to improve humanity [14].

The COVID-19 pandemic is, according to [15], more than a sanitary emergency. This is already a crisis with ramifications for the most vulnerable artists and professionals working in museums and exhibitions. Many states have sought to adopt virtual exhibitions during this time to mitigate losses caused by the closing of permanent physical exhibitions as a result of global societal restrictions. The benefits of virtualizing the interiors of museums to the audience in order to examine certain essential elements that they would not have observed with the naked eye within the structure are presented in [16], as well as the limitations of such a representation. One drawback is that when the public visits the museum, they will not experience the same experience as they would in the real location.

3.4 A Methodology for Assessing the Economic Potential Generated by Heritage Collections

One can assess how profitable it is to implement an exhibition before making it available to the public based on the business model used by virtual exhibition developers.

In this industry, there are a variety of business models that can be successfully implemented: The most relevant model is the *advertising revenue model*, which entails the placement of commercials in a virtual exhibition. A sponsor may provide financial support or products and services in the form of *corporate sponsorship. Freemium model* allows visitors to access the exhibition for free but charges a fee for additional features or services; The *donation-based crowdfunding* strategy entails obtaining funds from a community of users who do not expect a monetary return on their contribution [17].

From the perspective of the organization, business models depict how value is created, delivered, and captured. If business models are to be applied to generate money for public cultural organizations, they must be looked at in a larger context. The creation of new business models for the creative reuse of digital cultural content must: (a) enable for greater access to cultural heritage collections, and (b) produce cash to support cultural institutions' long-term viability.

Business models should allow for greater access to cultural resources while also protect copyright and intellectual property and generating revenue to ensure the long-term viability of third-party digital content exploiters. The use of modern ICTs in cultural institutions will help them advance culturally and commercially. Many elements of human activity have been drastically altered by the current technology revolution. While some of them have disappeared, other have been reinvented.

If a cultural institution decides to use modern ICT to create one or more virtual exhibitions, both as mobile and web applications, and uses appropriate marketing strategies to promote cultural heritage collections, the virtual exhibitions will generate visitors/virtual users who are likely to subsequently become real visitors to the cultural institution. Finally, a rise in the number of genuine visitors will boost the cultural institution's revenue.

When a cultural heritage organization chooses to introduce a new virtual exhibition, it must take into account the costs of hiring IT specialists as well as the price of acquiring certain software tools and licensing. Designing and implementing virtual exhibitions using ICT platforms and tools is a complex undertaking that necessitates a number of decisions at various stages. The cultural heritage institution or other sponsorships supplied by private corporations or non-governmental organizations should compensate these costs. There are a number of crucial aspects to address, both technical and non-technical. Multi-attribute decision criteria and models could be useful in resolving frequent decision situations [18].

Virtual exhibitions could incorporate interactive and virtual elements based on augmented reality to encourage visitors to participate and can be upgraded using *augmented reality mobile systems* [19] features to replace real-world things with simulated ones and deliver the optimum experience for the user. In the near future, the Internet of Things is expected to have a significant impact on the cultural sector's evolution. Adopting the most advanced technologies, such as the Internet of Things (IoT) and the Internet of Services (IoS), can help smart cities achieve unified ICT platforms for a variety of applications.

An expanding number of innovative services and increased customer demands are driving the rapid growth of the Information and Communications Technology industry today. In general, the number of interconnected handheld devices has been rapidly increasing year after year, thanks to both consumers and widespread adoption of the Internet of Things [20].

Smart cities of the future are expected to rely not only on sensors, but also on a vast number of gadgets that will integrate their data into smart platforms for analyzing people's and large communities' activities in a city. Important cultural heritage pieces, such as monuments, sculptures, and statues, can be fitted with sensors and "beacon" technology, allowing direct contact between the cultural object and visitors approaching it.

Culture is more valuable than ever in times of crisis. Culture is strengthening our societies' resilience in the face of unprecedented and difficult circumstances. The negative impacts of the COVID-19 crisis on the cultural and creative sectors should be mitigated.

Overall, given the strong evidence presented here on the impact of cultural and societal backgrounds on young people's educational and technological perspectives, an understanding of the potential effectiveness of any dispersed decision-oriented or design-oriented effort from the perspective of adaptation to the mentioned backgrounds is extremely valuable.

3.5 Using the Resource-Based View Concept

The concept of *resource-based view*, or RBV, denotes a strategy developed by businesses to better comprehend the various aspects of their operations in order to gain a long-term competitive edge [21, 22]. Businesses have grown more competitive as a result of technological breakthroughs and ever-growing innovations as a work culture. Today, every company's ultimate goal is to create value propositions for their customers in order to stay relevant in the face of ever shifting market trends. The extent to which strategy integration is implemented in present business functions provides a competitive advantage.

Managers in an RBV-centered organization should devise a strategy for using internal resources to take advantage of external opportunities and competition. When using a resource-based view of the company, the following are the processes to developing a strategy [21]:

- Determine the most important resources and skills.
- Assign qualified personnel to projects, pool resources for numerous projects, and so on.
- Implement niche skill succession planning.
- Regularly train the resource pool to improve their skills.

In the resource-based view, it is claimed that the value would be the only and necessary condition. It is also claimed that no resource or technique is useful in and of itself: it is dependent on a set of resources, processes, and integrated assets. In terms of the RBV attribute of imitation, we might question to what extent a precious asset is independent of other resources, and hence to what degree a resource combination is rare in and of itself. The creation of value and its embedding in a resource configuration is very important. The other primary conditions in the RBV could be challenged if the definition of value is revised: rarity and cost of copying, and inability of replacement with strategic equivalents [23].

According to the RBV concept, whether organizations will make higher profits and have a competitive advantage over others is determined by who owns and controls strategic assets. To determine the impact of resources, three major questions are asked:

- How valuable is the resource or capability?
- Is it dispersed differently among rival firms?

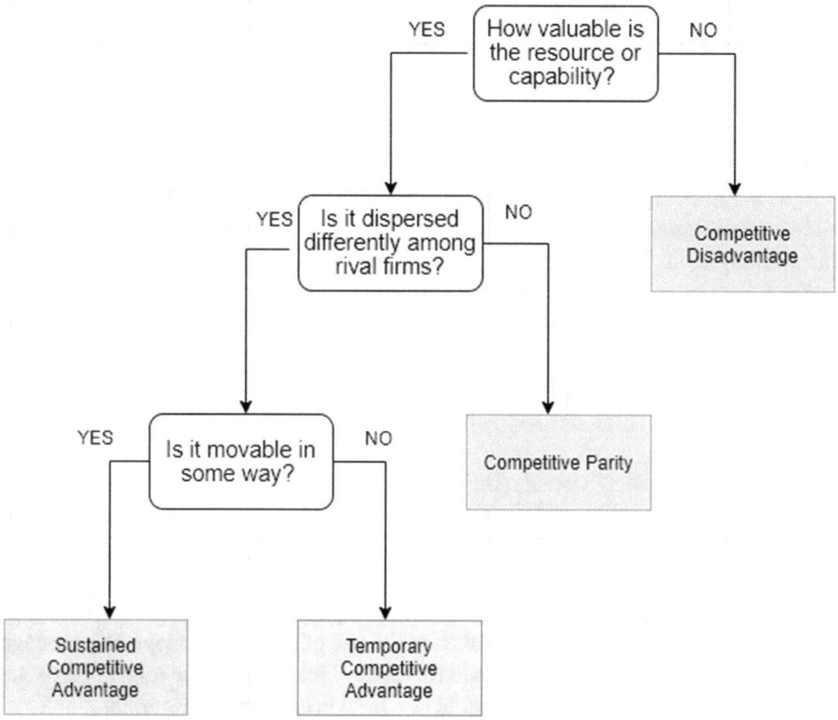

Fig. 3.4 Identification of resources and capabilities (adapted from [22])

- Is it movable in some way? The only way to obtain a long-term competitive advantage is to answer yes to all three questions [22].

As shown in Fig. 3.4, the problem of a resource's worth is usually answered in one of two ways. First, a resource might be considered valuable if it is used to minimize a company's costs (low-cost resources). Second, a resource might be considered as valuable if it is used to boost a sales profit (differentiated resources).

3.6 Conclusions

In these times of digital change, galleries, libraries, archives, museums and our entire attitude toward our cultural legacy are at a critical crossroads in a post-pandemic era [14, 24].

The cultural industry can benefit from the emergence of new technologies to boost the market value of cultural heritage collections, taking into consideration the impact of current ICT on all aspects of human activity.

At the same time, the cultural institutions should adapt themselves to the needs and behavioral-cultural profiles of visitors to cultural institutions. They should not be overly reliant on public budget financing, given the current economic position in Europe and future strategies, and should strive to become self-sufficient in this regard.

To meet the requirements of modern work, people need STEM (Science, Technology, Engineering, and Mathematics) abilities. We expect such new talents to enable remote work in the near future through the use of current ICT. We are currently transitioning from the industry 4.0 paradigm to the *Society 5.0* paradigm, which is characterized by rapid technological advancement [25].

Virtual exhibitions are the one of the most effective approaches to tell the story of each cultural object through the use of modern technology. A virtual exhibition today has a significant obligation to supply people with information, knowledge, and instructional content, and it is a sensible answer for cultural organizations' long-term sustainability.

In the coming years, virtual exhibitions will see a considerable increase in the use of online and mobile applications. They provide a number of advantages over traditional exhibitions, including the ability to display digital replicas of tangible cultural heritage objects.

Given the novelty of the research topic, developments in information technology, and the involvement of European bodies in the use of cultural heritage as a strategic economic resource, generating revenue for cultural institutions and related fields, the relevance of future research openings in this field is especially important.

References

1. Jones, L., Aho, M.: Resilience by design. Bridge **49**(2) (2019). https://www.nae.edu/212175/Resilience-by-Design
2. McKinsey: Agile resilience in the UK: lessons from COVID-19 for the 'next normal, October 13. https://www.mckinsey.com/business-functions/organization/our-insights/agile-resilience-in-the-uk-lessons-from-covid-19-for-the-next-normal
3. Ciurea, C., Pocatilu, L., Filip, F.G.: Using modern information and communication technologies to support the access to cultural values. J. Syst. Manag. Sci. **10**(2), 1–20 (2020)
4. Ernst, Young: Rebuilding Europe, The cultural and creative economy before and after the COVID-19 crisis, January (2021). https://assets.ey.com/content/dam/ey-sites/ey-com/fr_fr/topics/government-and-public-sector/panorama-europeen-des-industries-culturelles-et-creatives/ey-panorama-des-icc-2021.pdf
5. Ciurea, C., Filip, F.G.: The globalization impact on creative industries and cultural heritage: a case study. Creat. Stud. **12**(2), 211–223 (2019)
6. Visan, M., Ciurea, C.: Smart city: concepts and two relevant components. Int. J. Comput. Commun. Control (IJCCC) **15**(2), 1–9 (2020)
7. Borda, A., Bowen, J.P.: Smart cities and digital culture: models of innovation. In: Giannini, T., Bowen, J. (eds.) Museums and Digital Culture. Springer Series on Cultural Computing. Springer, Cham (2019)
8. Filip, F.G., Ciurea, C., Dragomirescu, H., Ivan, I.: Cultural heritage and modern information and communication technologies. Technol. Econ. Dev. Econ. **21**(3), 441–459 (2015)

9. Dower, N., Williams, J. (Eds.): Global Citizenship: A Critical Introduction, 1st ed. Routledge (2002). https://doi.org/10.4324/9781315023595
10. Ciurea, C., Filip, F.G.: Virtual exhibitions in cultural institutions: useful applications of informatics in a knowledge-based society. Stud. Inform. Control **28**(1), 55–64 (2019)
11. Coskun, V., Ozdenizci, B., Ok, K.: A survey on near field communication (NFC) technology. Wirel. Pers. Commun. **71**, 2259–2294 (2013). https://doi.org/10.1007/s11277-012-0935-5
12. Russo Spena, T., Bifulco, F.: Digital Transformation in the Cultural Heritage Sector, Challenges to Marketing in the New Digital Era. Springer, Cham (2021). https://doi.org/10.1007/978-3-030-63376-9
13. Hofkirchner, W.: Digital humanism: epistemological, ontological and praxiological foundations. In: Verdegem, P. (ed.) AI for Everyone? Critical Perspectives, pp. 33–47. University of Westminster Press, London (2021). https://doi.org/10.16997/book55.c
14. Eichinger, A., Prager, K.: We are needed more than ever: cultural heritage, libraries and archives. DIGHUM—perspectives on digital humanism. https://dighum.ec.tuwien.ac.at/perspectives-on-digital-humanism/we-are-needed-more-than-ever-cultural-heritage-libraries-and-archives/
15. Cobley, J., Gaimster, D., So, S., Gorbey, K., Arnold, K., Poulot, D., Jiang, M.: Museums in the pandemic: a survey of responses on the current crisis. Museum Worlds **8**(1), 111–134 (2020)
16. Vajda, A.: Museums and online spaces. The society-building role of the museums during the pandemic. Acta Univ. Sapientiae Commun. **7**, 42–53 (2020)
17. Ciurea, C., Filip, F.G.: Identifying business models for re-use of cultural objects by using modern ICT tools. Inform. Econ. **22**(1), 68–75 (2018)
18. Filip, F.G., Zamfirescu, C.B., Ciurea, C.: Computer-Supported Collaborative Decision-Making. Series Automation, Collaboration, & E-Services, 216 pp. Springer (2017). ISBN 978-3-319-47219-5
19. Carmigniani, J., Furht, B., Anisetti, M., et al.: Augmented reality technologies, systems and applications. Multimed. Tools Appl. **51**, 341–377 (2011). https://doi.org/10.1007/s11042-010-0660-6
20. Ometov, A., Shubina, V., Klus, L., Skibińska, J., Saafi, S., Pascacio, P., Flueratoru, L., Quezada Gaibor, D., Chukhno, N., Chukhno, O., Ali, A., Channa, A., Svertoka, E., Bin Qaim, W., Casanova-Marqués, R., Holcer, S., Torres-Sospedra, J., Casteleyn, S., Ruggeri, G., Araniti, G., Burget, R., Hosek, J., Lohan, E.S.: A survey on wearable technology: history, state-of-the-art and current challenges. Comput. Netw. **193**(108074) (2021)
21. Yesodharan, S., Mohan, N.: Using the resource-based view strategy for a competitive advantage, March 4, 2021. https://www.saviom.com/blog/using-the-resource-based-view-strategy-for-competitive-advantage/
22. Madhani, P.M.: Resource based view (RBV) of competitive advantage an overview (March 26, 2010). In: Madhani, P. (ed.) Resource Based View: Concepts and Practices, pp. 3–22. Icfai University Press, Hyderabad (2009). Available at SSRN https://ssrn.com/abstract=1578704
23. Salazar, L.A.L.: The resource-based view and the concept of value the role of emergence in value creation. Mercados y Negocios **35**, 27–46 (2017)
24. Palumbo, R.: Enhancing museums' attractiveness through digitization: an investigation of Italian medium and large-sized museums and cultural institutions. Int. J. Tour. Res. 1–14 (2021). https://doi.org/10.1002/jtr.2494
25. Pereira, A.G., Lima, T.M., Charrua-Santos, F.: Industry 4.0 and society 5.0: opportunities and threats. Int. J. Recent Technol. Eng. (IJRTE) **8**(5) (2020)

Chapter 4
eTeaching and eLearning Resources. A Challenge for University Education During Covid-19

Larisa Ivascu and Alin Artene

Abstract Concepts such as eLearning and eTeaching have become a reality today in a pandemic context facing people around the world. eTeaching and eLearning enables teachers and students to take online courses with the help of electronic equipment such as computers, laptops, tablets and even smartphones. Despite the shortcomings of adapting traditional university courses to online teaching materials, eLearning and eTeaching are effective methods of education, monitoring the academic development of students and finally yet importantly assessing the knowledge of students in the university environment. Another aspect worth considering is the reduction of costs in terms of education because when the courses went online the expenses necessary for the schooling and accommodation of students represented a fraction of their cost in the pre-pandemic period. In this chapter, we will analyze the challenges that universities, professors and students have had to overcome in order to build a reliable and efficient electronic educational system.

Keywords eTeaching · eLearning · Efficient electronic educational system · European digital decade · Autonomous learning system · Digitally responsible Europe · Technology based learning · Open distance learning · Distributed learning

4.1 Introduction

Since the beginning of the global pandemic of COVID-19 in Europe, with its first case confirmed in France, on January 24, 2020, the need for eLearning and eTeaching

L. Ivascu · A. Artene (✉)
Faculty of Management in Production and Transportation, Politehnica University of Timisoara, Timisoara, Romania
e-mail: alin.artene@upt.ro

L. Ivascu
e-mail: larisa.ivascu@upt.ro

L. Ivascu
Academy of Romanian Scientists, Ilfov 3, 050044 Bucharest, Romania

L. Ivascu et al. (eds.), *Intelligent Techniques for Efficient Use of Valuable Resources*, Intelligent Systems Reference Library 227,
https://doi.org/10.1007/978-3-031-09928-1_4

began to take shape. Due to the COVID-19 pandemic, a series of non-pharmaceutical interventions known colloquially as lockdowns have led to the closure of higher education institutions all around Europe, creating challenges at all levels of education, especially for university students and professors [1, 2].

Currently, the European university environment uses the eLearning and eTeaching platform, which in most cases, is based on the Moodle-LMS platform or on some collaborative educational platforms designed on a global scale such as Zoom or Google Classroom, which incorporates mail modules, classroom and virtual rooms, video conferencing, presentations, and countless opportunities for testing and remote working [3–7].

By choosing a university, young students are far from home and family, they are more often faced with the pressure of making friends to be socially accepted, they have to make the transition to university or master's studies. All this had an amplified impact on those in the university environment during the pandemic.

This prolonged period of uncertainty has been a challenge for universities to approach things in the most correct and coherent way possible.

In addition, although the university environment has proven its ability to adapt to the new present, there are clear lessons to be learned from how the pandemic was handled in academia. In addition to access to quality eLearning, universities should consider the mental health of the beneficiaries of the teaching act and consider ways to support their university experience in advance. eLearning and eTeaching have developed worldwide in recent years, although these forms of education were never considered part of the majority formal education of the institutions until the spread of Covid-19.

Universities using university autonomy have adapted, according to their resources and ingenuity, to accommodate new opportunities for education systems, especially in the integration of technologies in learning and teaching Adapting the university environment to eLearning and eTeaching has led to ease of use, flexibility and better control over the academic environment.

For university professors, the COVID-19 pandemic period was an adaptive and transformative challenge par excellence, one for which there was no user guide or set of good practices and to which they connected and adapted on the fly.

The resources available to universities, depending on their purpose, included material resources such as tangible and intangible assets, curricular resources, and a solid human resources strategy.

Curriculum and curricular resources included media content such as courses and seminars, videos, interactive modules, webinars, and other resources that directly support students in accumulating the knowledge and skills needed for the student's chosen curricula.

In terms of staffing strategy, the resources available to universities were those that supported teachers and assistants by directing them to media content that contributed to the development of their eTeaching skills.

Finally yet importantly, the material base, including tangible and intangible assets, was the one that supported the whole framework and included tools that helped to manage the teaching and assimilation of information, such as online communication tools, didactic management systems and any other type of devices or tools meant to create digital academic content.

4.2 Defining eLearning and eTeaching

Learning is the process that determines a selective, permanent and oriented change in a certain direction in knowledge and behavior. The vision of the learning process is included in the paradigm of student-centered education, the development of skills and the use of empirical evidence. This general framework of learning in the university environment is complemented by other values such as collaboration, reflexivity, and involvement.

An effective learning process is one in which students meet the educational objectives embodied in expected learning outcomes, developing their professional and transversal skills aimed at applying learning principles such as collaboration, contextualization, self-direction, quantifying teaching and assessment efforts.

E-learning is a current way of developing education in close connection and dependence on technological development and is synonymous with Online Learning, Web Based Learning—WBT, Internet Based Learning, Technology Based Learning, Open Distance Learning, Distributed Learning.

According to The Economic Times 2022 eLearning is defined as a learning system based on formalized teaching but with the help of electronic resources is known as E-learning. E-learning can also be termed as a network enabled transfer of skills and knowledge, and the delivery of education is made to a large number of recipients at the same or different times. The rapid evolution of technology and the progress in learning platforms and systems, contributed to the popularization of this form of education.

The European Commission through its Knowledge Centre on Interpretation 2022 considers "E-learning is a self-paced online learning method that uses new multimedia technologies for training. Using the Internet improves the quality of learning, as it facilitates access to resources and services, as well as exchanges and distance learning. This method provides the learner with a multisensory learning experience (sound, graphics and interactivity) that allows a better understanding and assimilation of knowledge" [8].

In the European Union's vision of the near future, eLearning represents a key component of all actions aimed at achieving the Lisbon goals, supporting economic growth the quality of life and the quality of job offers. The process of eLearning in universities is a form of education that involves distance interaction between the participants, an interaction that is often asynchronous. The main difference between distance learning or part-time education and eLearning is that the last one has a high degree of interaction, the interaction-taking place on the following levels: between participants, participant—material, participant—teacher or assistant [9, 10].

eTeaching involves attracting students in learning by electronic means; Electronic teaching consists of teacher-student interaction in the active development of specialized skills. A university professor needs knowledge of the curriculum, technological limitations of students, in the context of COVID-19 pandemic, and how to turn them into active learners [11].

We can say that a good eTeaching requires a commitment to the systematic understanding of eLearning. The purpose of eTeaching is not only to transmit information, but also to digitally transform students into active recipients of knowledge. Teaching is fundamentally about creating the digital and electronic pedagogical, social and ethical conditions of the future in which students agree to take over the leadership of their own learning, individually and collectively in the online environment [12, 13].

eTeaching correlated the approach of university teachers with that of students include and online tools were meant to have this presence in academia. eTeaching in the university environment has integrated several technologies and contributed to the improvement of the learning as well as the teaching experience [14].

In Universities, learning resources have evolved with new technologies, the learning process having to keep up with these new technologies. eTeaching has had effects on student retention and academic success rates [15–17].

The concern to deliver a quality teaching activity, using electronic means has led to its theorizing. Thus, in terms of content, eTeaching activity is complex. However, the same methods do not apply to the whole eLearning process. There are methodological differences between eLearning and eTeaching in the new complex social reality, that of the post-Covid-19 era, the same methods are used by the teacher to teach and the students to assimilate the information provided. Being a task of the teacher, the methodological elaboration of the electronic teaching contains problems related to its transformation into training. eTeaching and eLearning are complex and complementary; their reality is summarized in Table 4.1.

The university reality of e-learning includes a variety of e-learning tools such as LMS platforms, webinars, online assessment tools, management tools for university training, social interaction platforms, tools and information exchange between teachers and students, mobile devices and online platforms and micro e-learning, augmented reality, and virtual reality in the service of laboratory applications and seminars. From a university perspective, eLearning and eTeaching activities reduce the time of use and management of training: from the registration of users on the virtual campuses of universities, to the distributed results and the administration of assessment exams, reduces costs, cre4ats updated content and combines traditional and online practices as followed [6, 7, 9, 12, 18, 19]:

a. Tailored education, specifically to the student needs; E-learning offers university teaching activities advantages over traditional teaching by eliminating the time and attendance constraints for the student, which leads to a saving of time dedicated to training, thanks to the possibility to use the content at any time and from any place when the student needs being always available online.
b. Reducing costs with traditional teaching activities e-learning allows universities to reduce many important expenses, such as: material costs due to virtualization

Table 4.1 The differences between eTeaching and eLearning

eTeaching characteristics	eLearning characteristics
eTeaching must be an effective synchronous, asynchronous, or mixed interaction between teacher and student	eLearning addresses anyone, anywhere, anytime
More than traditional teaching, it requires exercise, techniques, procedures and IT skills and creativity	eLearning should be user-friendly and easy to use
eTeaching is dominated by the skill of communication and online social interactions	eLearning involves personalized learning and flexible pace
eTeaching should be well planned and the teacher should decide the objectives, synchronous or asynchronous methods of online teaching and assessment techniques	eLearning modules are often cheaper than traditional physical university courses
eTeaching is an online interaction activity with the organization, management, creation of quality online content and evaluation of results	eLearning has a lower impact on the environment
eTeaching should focus on eliminating anxiety caused by isolation and ensuring emotional stability for students	eLearning alternates sequential media content and invites the student to discussions, debates and personal initiatives
eTeaching stimulates students' thinking power and leads them to self-learning	eLearning uses inclusive language, being student-centered, and online content speaks directly to the student, rather than addressing the learning audience as a group
Teaching combined with Big Data Analysis can be observed, analyzed and evaluated	eLearning offers opportunities for self-reflection as students are encouraged to link information and the benefits of this information
eTeaching is a specialized task and can be taken as a set of component skills to achieve a specified set of objective instruction	eLearning gives students opportunities to correlate content with their personal role or situation through self-reflection
eTeaching allows the university to personalize the learning experience by selecting a visual theme or hosting individual comfort and accessibility	eLearning allows for easy navigation and is easy to use, allowing students to take control and quickly find what they need
	eLearning meets individual needs. based on existing skills

of work tools, elimination of costs associated with replacing staff in training, elimination of travel costs by improving the monitoring and management system

c. Track, evaluate and improve university activity with the possibility to customize the courses or adapt them to the needs of small groups, thus obtaining continuous system improvement.
d. Optimized and updated content with precise and exhaustive academic content and integrated with in-depth materials that allow university staff to manage

each content independently and to intervene in a targeted and fast way without affecting the whole project

e. eLearning and eTeaching system aimed at mobile learning and micro-learning can be carried out in any context and environment resulting in a student-centered teaching strategy, in which the reduced content is provided and synthesized to reduce cognitive overload.
f. Involving students in the use of game-related dynamics applied to learning to facilitate student participation and simplify the content of application activities and seminars.
g. The inclusion of Big Data learning analytics in training, or the collection of information generated by students in the course, plays a key role in predicting and improving the academic information provided.
h. Using augmented reality or virtual reality technologies, conferring application activities and seminars, an immersive experience, supported by artificial intelligence, and allowing students to face potential real situations, protected in a "secure" simulation.

4.3 Type of eLearning and eTeaching

In European universities, eLearning or eTeachings are forms of education characterized by the physical separation of teachers and students in the training process and the use of diversified technologies to facilitate student teacher and student–student communication. The use of different technologies can be carried out remotely in a synchronous or asynchronous manner, using various equipment connected to the internet [7, 12, 17].

Synchronous courses are courses in which the partners of the university teaching act participate at the same time, but in separate locations, generally other than the university campus. Synchronous eTeaching involves a structured learning strategy, in which courses and seminars are scheduled at pre-set times, address study groups called virtual classes, and students benefit from real-time interactions through platforms such as Zoom or Google Meet [4, 14, 20, 21]. Synchronous eTeaching occurs when students and teachers communicate in real time. Classroom teaching, classroom and seminar seminars, smartphone, video conferencing, and social media chat are examples of synchronous communication. There are a lot of tools universities can use for video and audio university lectures and according to Okta 2020 Bussines at work report Zoom is the top video conference app, Fig. 4.1.

Another platform useful in the university environment is a chat platform, Fig. 4.2, which confer users the ability to scroll back and forward trough the chat and find anything they might have missed.

Asynchronous eTeaching are courses in which students do not participate at the same time as teachers, and academic open access media content to which students have access through cloud services. This form of eLearning involves individual study, in print or digital format, open access, implemented virtual campuses of universities.

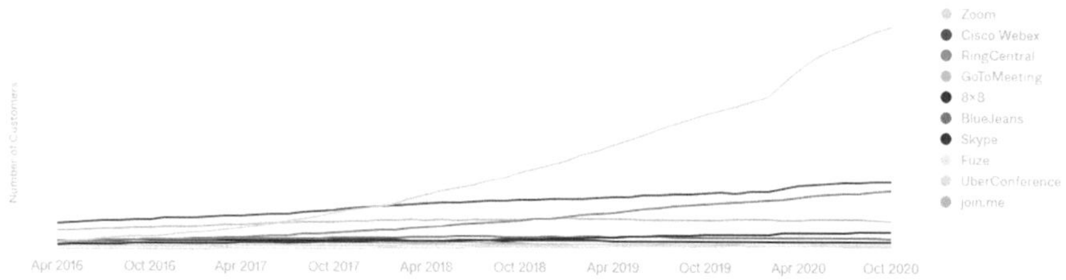

Fig. 4.1 Video conferencing apps. *Source* https://www.okta.com/businesses-at-work/2021/

Fig. 4.2 Popular messaging apps worldwide. *Source* https://www.similarweb.com/corp/blog/research/market-research/worldwide-messaging-apps/

Asynchronous eTeaching is more flexible than synchronous eTeaching. Courses and application activities take place at one time and are stored in the Cloud for students to access at another time, whenever it is most convenient for the students [14, 22].

In the early days of online eTeaching activities, most courses in universities were asynchronous, an updated version of distance learning or low attendance learning. In the Covid-19 pandemic period universities upgraded to virtual classroom management systems and audio tools that allow university staff and students to do almost everything originally occurring in classrooms on campus. Additionally, universities

Table 4.2 SWOT analysis for university eTeaching and eLearning

	Helpful for eTeaching and eLearning	Harmful for eTeaching and eLearning
Internal	Strength Flexibility in teaching and autonomy Management and independence in the university campus Universities can link various resources in several varying formats Direct assistance in the teaching and learning process Provides detailed reports Low education costs Offers self-paced learning and comfort Fast courses delivery Dynamic and adaptable learning process Offers traceable outcomes for university program beneficiaries	Weakness eLearning has created a socioeconomic divide of students due to the lack of access to resources No social and interpersonal interaction with eLearning Because universities have more complex skills and competencies, it is more difficult to put together an effective eLearning program eLearning materials can become outdated Investments in tangible and intangible assets Only 50% of professors approve of online learning Not most professors see eLearning as being as effective as in-classroom instruction
External	Opportunity The eLearning market is expected to surpass 370 billion dollars by 2026 In Covid-19 Pandemic, almost every university moved classes online Majority of universities have purchased additional IT technology for students The majority of universities offered learning opportunities via mobile devices By next year, eLearning management systems will be valued at 23 billion dollars	Threat Protecting intellectual property rights In eLearning For universities eLearning and eTeaching takes time, monetary resources, and a lot of expertise. A good eLearning course involves multimedia, custom web development, technical support, and strong User Interaction (UI) design If a business model changes or market conditions are disrupted, university eLearning can even become obsolete Copyrights in eLearning

trough their virtual campuses engage students in collaborative and reflective activities stored in cloud or granted trough open access sources. Table 4.2 shows a SWOT analysis for university eTeaching and eLearning.

4.4 Techniques and Platforms Used in eTeaching and eLearning in Universities

We can't help but wonder what the universities will do with the available technology and the know-how acquired after the Covid-19 pandemic. Despite all the efforts of the participants in the academic life around the virtual or real university campuses, eLearning and eTeaching have not fulfilled their potential to coagulate the relationship between teachers and students. Soon, if we want to reap the benefits of eTeaching, universities will need to develop a solution for unversity and master and PhD education that is based on the relationship between student and university professors, and not just on the reproduction of eTeaching materials available [22–24].

In this scenario, the role of university professors is essential. The solutions to successful eTeaching are competent, paid teachers who can provide help and support, as well as transfer of material for eLearning, combined with quality academic content and IT technology designed to support university teaching. In turn, universities will need to identify new ways to attract sufficient funding for the sustainability of eTeaching and eLearning as a new reality, parallel to traditional teaching and learning methods. It is also the responsibility of universities to protect eLearning content and to protect the rights of eTeaching authors in an open access academic environment. The new pandemic reality strongly reminds us that universities rely heavily on the interaction between professors and their students, students, and colleagues, as well as professors with their colleagues and collaborators [9, 25–28].

The Covid-19 pandemic demonstrates to universities what abilities and social skills students need in this unpredictable and ever-changing world, such as sustainable mindsets, decision-making, creative problem-solving, and perhaps, above all, adaptability to new complex realities. eTeaching and eLearning can include acquiring the most reliable methods that universities have developed to guide their students to discover the truth about themselves and the world in the exercise of their skills [1, 22].

When university programs provide essential skills to graduates, teachers, through eTeaching methods and techniques, make the student aware of their assets and through eLearnig platforms encourages them to be intelligent and reflective in exercising their skills. An electronic university platform for eLearning must independently manage the entire curriculum of academic courses and seminars, whether it is a theoretical approach or a new trends and innovations in the fields of studies, or application activities and seminars. An eLearning platform should not replace the human teacher-student interaction necessary for professional development but should witness the exponential expansion of eLearning skills. Universities need to allow students to study some topics of interest entirely online and remotely [11, 13, 20, 26, 29].

Respecting and not compromising the integrity and autonomy of the university, each higher education institution will structure differently their electronic courses and seminars, but the agency responsible for quality assurance will follow and include some common elements to follow through the university eLearning platforms [15].

The challenges for universities in the eLearning industry are high, because according to Forbes, the global eLearning market is estimated to generate 325 billion dollars in revenues until 2025 and the number of people having and using smartphones is 6.64 billion or 84% of the world population according to bankmycell 2022, Table 4.3.

As interconnectivity and the new reality offer new opportunities for individuals to gain easier access to platforms for training and learning tools, universities must also compete with these service providers. Because in a recent study by Statista, 20% of students worldwide had taken an online course in the preceding 12 months, we

Table 4.3 Best eLearning platforms in 2022

Skillshare	27.000 courses
Mindvalley	50 courses
Brilliant.org	60 courses
LinkedIn learning	16.000 courses
MasterClass	80 courses
Udemy	150.000 courses
Edx.org	2.500 courses
Udacity	240 courses
Coursera	4.300 courses
Futurelearn	820 courses

Source https://sites.google.com/site/videoblocksreview/online-learning-platforms

present in the next table the most popular e-learning platforms that force universities to adapt their curriculum to new realities to tap in the multibillion-dollar market available for education soon [10, 14].

University management with the help of teachers and students engaged in decision-making process must transform and rethink existing traditional courses in eLearning material [9, 12, 18].

First, the university must analyze the profile of the students admitted to its bachelor and master's programs, analyze the learning environment available to them and review the learning objectives.

A special task belongs to university professors who this time find themselves in the position of electronic content creators. They must analyze the content of the courses, identify, and replace the content gap, and provide a summary and a summative assessment. Once the eTeaching objectives are identified and the strategy is established, teachers and students must decide on the level of interaction, the amount of audio and video materials [19].

These materials can be grouped according to the recipient and the specific objectives of the syllabus, in lessons on request, suitable both for individual study and for series and groups of students, live courses, which require a large transmission capacity, material for micro learning—especially for the lower years of the undergraduate cycle. In university eTeaching, teachers will be able to use the entire tools specific to micro learning accordingly: animations, info graphics, quiz, videos, mobile applications, concept maps, logical schemes and so on.

A positive aspect of the Covid-19 pandemic was that many personalized interactive learning platforms did not use the whole situation to earn extra money, but on the contrary, helped to overcome the crisis. Unlike most LMSs, Moodle is a free, open-source learning management platform. Moodle includes a wide range of drag and drop tools and useful resources that can help teachers and students and is used as an eTaching and eLearning platform for approximately 60 million online education sites.

Table 4.4 Popularity and functions of eLearning tools in 2022

Platforms	Zoom	Microsoft Teams	Google Meet	Cisco Webex	Microsoft Lync
	Skype	Vedamo	Miro	Google Classroom	Adobe Connect
Activities	Live sessions	Breakout rooms	Screen sharing	Annotations	Chat rooms
	Worksheets	Whiteboard	Raise Hands	Polls	Express emotions

In universities, the eTeaching platform that university teaching staff chooses depends on a variety of factors, including the size of the series and groups of students, the subject of the course, the needs of the group, the location of the participants, or the theoretical or applied nature of the course. Universities recommend that teachers should use a single platform so that they and students become familiar with it and the eLearning process is effective.

Analyzing the trends and popularity of the platforms among professors and students in universities, we present below the main platforms and the activities that can be carried out within them, Table 4.4.

4.5 Conclusions

Universities must build the infrastructure and make it accessible to all beneficiaries of bachelor, masters, and doctoral degree programs. Also, through the asset acquisition strategy to provide departments with a basic element in ITC. In terms of human resource strategy, universities need to invest in training university teachers on the use of technology, popularizing the use of digital technology and methods in eTeaching in all departments, and creating platforms with easier access for all users. Regarding ensuring the quality of education, universities need to engage in the development of plans to raise awareness among students about the benefits of eLearnig and the incorporation of technology as an additional learning tool.

eTeaching and eLearnig approach in universities cannot and should not be a substitute for traditional teaching. Universities must use digital teaching as an alternative and a faithful ally to e-innovation and the efficiency of electronic resources, and eTeaching methods.

We must not forget that the University, in addition to being an educational environment, is also a training place that cannot ignore the dynamics of personal and social interaction that make it up to promote the construction of the person beyond teachers and students.

Universities need to protect students from digital division that could become extreme if access to educational content depended on the latest technologies because at the time of the Covid-19 pandemic, only 60% of the world's population had access to the Internet.

Universities will need to ensure that students and faculty staff have broadband access to the Internet while ensuring equitable access to the internet within university campus area. University management will also need to provide teaching staff, students, electronic devices for access to the internet and software as well as adequate resources for research, communication, multimedia content creation and collaboration for use in and out of school.

References

1. Ayebi-Arthur, K.: E-learning, resilience, and change in higher education: helping a university cope after a natural disaster. E-Learn. Digit. Media **14**(5), 259–274 (2017). https://doi.org/10.1177/2042753017751712
2. Affouneh, S., Salha, S.N., Khlaif, Z.: Designing quality e-learning environments for emergency remote teaching in coronavirus crisis. Interdiscip. J. Virtual Learn. Med. Sci. **11**(2), 1–3 (2020)
3. Basilaia, G., Dgebuadze, M., Kantaria, M., Chokhonelidze, G.: Replacing the classic learning form at universities as an immediate response to the COVID-19 virus infection in Georgia. Int. J. Res. Appl. Sci. Eng. Technol. **8**(III) (2020)
4. Barrett, S.: Coronavirus on campus: college students scramble to solve food insecurity and housing challenges. CNBC (2020). https://www.cnbc.com/2020/03/23/coronaviruson-campus-students-face-food-insecurity-housing-crunch.html
5. Bayham, J., Fenichel, E.P.: The impact of school closure for COVID-19 on the US healthcare workforce and the net mortality effect (2020). medRxiv. https://doi.org/10.1101/2020.03.09.20033415
6. Bennett, B., Spencer, D., Bergmann, J., Cockrum, T., Musallam, R., Sams, A., Fisch, K., Overmyer, J.: The flipped classroom manifest. Int. J. Educ. Res. **4**(11), 1–8 (2013); Bjorklund, A., Salvanes, K.: Education and family background: mechanisms and policies. Handb. Econ. Educ. **3**(1), 201–247 (2011)
7. Bankmycell, How many smartphones are in the world? January 20201. https://www.bankmycell.com/blog/how-many-phones-are-in-the-world. Accessed 12 Jan 2022
8. https://ec.europa.eu/education/knowledge-centre-interpretation/content/2022-2023_ro. Accessed 1 Feb 2022
9. Brooks, S., Smith, L., Webster, R., Weston, D., Woodland, L., Hall, I., Rubin, J.: The impact of unplanned school closure on children's social contact. Eurosurveillance **25**(13), 21–31 (2020). https://doi.org/10.31219/osf.io/2txsr
10. Checa, D., Bustillo, A.: A review of immersive virtual reality serious games to enhance learning and training. Multimed. Tools Appl. **79** (2020). https://doi.org/10.1007/s11042-019-08348-9
11. Keeton, M.T.: Best online instructional practices: report of phase I of an ongoing study. J. Asynchron. Learn. Netw. **8**(2), 75–100 (2004)
12. Brianna, D., Derrian, R., Hunter, H., Kerra, B., Nancy, C.: Using EdTech to enhance learning. Int. J. Whole Child **4**(2), 57–63 (2019)
13. Kebritchi, M., Lipschuetz, A., Santiague, L.: Issues and challenges for teaching successful online courses in higher education. J. Educ. Technol. Syst. **46**(1), 4–29 (2017)
14. Choosing the right platform for live online teaching, UCL, United Kingdom (2022). https://www.ucl.ac.uk/teaching-learning/education-planning-2021-22/online-teaching-guidance-tips-and-platforms/choosing-right-platform-live. Accessed 12 Jan 2022
15. Dlearn: 5G technology and its influence on education (2019). https://dlearn.eu/5g-technology-and-its-influence-on-education/

16. The Economic Times, English Edition. https://economictimes.indiatimes.com/definition/e-learning. Accessed 10 Jan 2022
17. UNESCO: How to plan distance learning solutions during temporary schools closures (2020). https://en.unesco.org/news/how-plan-distance-learning-solutions-during-temporaryschools-closures
18. Briggs, B.: Education under attack and battered by natural disasters in 2018. TheirWorld (2018). https://theirworld.org/
19. Cluver, L., Lachman, J.M., Sherr, L., Wessels, I., Krug, E., Rakotomalala, S., Blight, S., Hillis, S., Bachman, G., Green, O., Butchart, A., Tomlinson, M., Ward, C.L., Doubt, J., McDonald, K.: Parenting in a time of COVID-19. Lancet **395**(10231), e64 (2020). https://doi.org/10.1016/S0140-6736(20)30736-4
20. Kim, K.-J., Bonk, C.J.: The future of online teaching and learning in higher education: the survey says. Educause Q. **4**, 22–30 (2006)
21. Piopiunik, M., Schwerdt, G., Simon, L., Woessman, L.: Skills, signals, and employability: an experimental investigation. Eur. Econ. Rev. **123**(1) (2020). https://doi.org/10.1016/j.euroecorev.2020.103374
22. Kawano, S., Kakehashi, M.: Substantial impact of school closure on the transmission dynamics during the pandemic Flu H1N1-2009 in Oita, Japan. PLOS ONE **10**(12), e0144839 (2015). https://doi.org/10.1371/014483919326203
23. Andrade, M.S.: Effective eLearning and eTeaching—a theoretical model. In: Gradinarova, B. (ed.) E-Learning—Instructional Design, Organizational Strategy and Management. IntechOpen (2015). https://doi.org/10.5772/60578. https://www.intechopen.com/chapters/48924
24. Okta 2020 Business at work. https://www.okta.com/businesses-at-work/2021/. Accessed 11 Jan 2022
25. Liguori, E.W., Winkler, C.: From offline to online: challenges and opportunities for entrepreneurship education following the COVID-19 pandemic. Entrep. Educ. Pedag. (2020). https://doi.org/10.1177/2515127420916738
26. Lee, J., Lubienski, C.: The impact of school closures on equity of access in Chicago. Educ. Urban Soc. **49**(1), 53–80 (2017). https://doi.org/10.1177/0013124516630601
27. Littlefield, J.: The difference between synchronous and asynchronous distance learning (2018). https://www.thoughtco.com/synchronous-distance-learning-asynchronous-distance-learning-1097959
28. Martin, A.: How to optimize online learning in the age of coronavirus (COVID-19): a 5-point guide for educators (2020). https://www.researchgate.net/publication/339944395_How_to_Optimize_Online_Learning_in_the_Age_of_Coronavirus_COVID-19_A_5-Point_Guide_for_Educators
29. European Commission, Knowledge Centre on Interpretation (2022). https://ec.europa.eu/education/knowledge-centre-interpretation/conference-interpreting/professional-support/e-learning-and-online-resources_ro. Accessed 10 Jan 2022

Chapter 5
Challenges and Opportunities for the E-Learning Users During COVID-19 Pandemic Times

Aura Emanuela Domil, Nicolae Bobițan, Diana Dumitrescu, Valentin Burcă, and Oana Bogdan

Abstract The restrictive measures, taken worldwide, for the COVID-19 pandemic transferred teaching activities into the online environment. These circumstances have forced the restructuring of the face-to-face teaching process. We focused our research on measuring student's perception, through a questionnaire distributed online. Our survey focused on the changes of the learning process, during COVID-19 pandemic, and how teachers have reacted to these changes to overcome students' limitations in online learning. This chapter explores in the first part the exploratory insights within the literature, and in the second part, with empirical evidence which highlights those students prefer the hybrid teaching system, and that teachers have managed to overcome the barriers/obstacles generated by the transition from traditional to the online teaching system.

Keywords Online learning · Technological based learning resources · Learning and teaching tools · Multinomial regression · Categorical principal components analysis · Teaching and learning resilience

---

A. E. Domil (✉) · N. Bobițan · D. Dumitrescu · V. Burcă · O. Bogdan
Faculty of Economics and Business Administration, West University of Timisoara, Timisoara, Romania
e-mail: aura.domil@e-uvt.ro

N. Bobițan
e-mail: nicolae.bobitan@e-uvt.ro

D. Dumitrescu
e-mail: diana.dumitrescu@e-uvt.ro

V. Burcă
e-mail: valentin.burca@e-uvt.ro

O. Bogdan
e-mail: oana.bogdan@e-uvt.ro

L. Ivascu et al. (eds.), *Intelligent Techniques for Efficient Use of Valuable Resources*, Intelligent Systems Reference Library 227,
https://doi.org/10.1007/978-3-031-09928-1_5

5.1 Introduction

Throughout history mankind has faced a series of pandemics that have unbalanced economies and triggered political and social change through a domino effect.

According to the World Health Organization (WHO) the Justinian plague, recorded around 541–542, has killed more than 100 million people. The bubonic plague or "black death of the fourteenth century" killed about 50 million people on 3 continents, namely Asia, Africa and Europe [5].

Cholera, Yellow Fever, Typhoid, Smallpox, Spanish Flu, Polio, MERS and SARS have also caused countless infections and deaths, and some of them still cause thousands of deaths every day.

Starting from the premise that history repeats itself and globalization has accelerated the spread of viruses in the world, nowadays we face a new pandemic, respectively a new strain of coronavirus, SARS-CoV2, which according to the United Nations determined from an economic point of view an unprecedented impact since the Great Recession [27].

Due to the rapid spread and the extremely large number of victims requiring hospitalization and specific treatment, on March 11, 2020, the WHO classified COVID-19 as a pandemic, namely a widespread epidemic beyond international borders, affecting a large number of people [7]. Thus, governments around the world have focused their attention and allocated significant sums to introduce restrictive measures to combat both the spread and the increase in the number of infections. Social distancing, washing hands with soap and water, wearing a mask, quarantining and closing areas of activity where the spread of the virus cannot be controlled, such as hotels, restaurants, bars, sports and cultural activities are examples of measures taken to combat the spread of the virus.

Teaching activity has also been affected, being transposed into the online environment. Both teachers and students were forced to adapt to new teaching and learning methods, almost overnight, in order to deal with the "new normal" created by the coronavirus disease and provide, in the same time, a quality education.

School closures to mitigate the spread of COVID-19 have affected nearly 1.6 billion learners in over 190 countries. According to the UNESCO Global monitoring database, in February 10, 2020, when not much was known about the novel coronavirus, 883.055 learners were affected. A few days after the pandemic was stated and governments imposed restrictive measures, namely on March 16, 2020, figures highlighted that 666,741,569 learners were affected by schools' closures. Nowadays, the latest figures released in 2022 point out a decrease to the value of 37,737,815 learners affected [30]. In Romania's case, all schools were closed from March 11, 2020 until the end of April and after, a hybrid teaching system was implemented in most of the Romanians schools.

Although it was considered just a backup educational system needed in order to continue the learning process in pandemic times [23], the impossibility to attend physical classes, the lack of digital skills [6] or adequate space and emerging technologies needed for online learning may affect, on the long term, school performance and the entire educational process [29]. Our aim in this paper is to bring some insights on the implications generated by the transition of the teaching activities into the online environment. In this context, we focused our research on measuring student's perception on how the novel coronavirus disease has change their way of learning and how teachers have reached to overcome the issues students' have to face due to online learning because of the COVID-19 pandemic restrictive measures, through a questionnaire distributed online.

This way we try to get an updated image of how the actual university online learning system is perceived by the final beneficiary, respectively the students, as on a global level there have been claimed several systemic issues that affect the quality of educational activities, such as: insufficient teachers' preparedness to support online learning, or ineffective communication between teachers and students in the context of transition to online teaching [25]. Additionally, we raise awareness through this study that a national resilience program dedicated for education system is essential to address students' learning losses, which should include clear action plans with precise due dates aimed to ensure higher flexibility from teachers' side and higher adaptability from students' side [28]. Otherwise, the current inequalities in education services will increase, especially in the rural area and disadvantages groups of students.

Our chapter comes with additional insights within the literature. It adds empirical evidence that highlights that students prefer the hybrid teaching system and that teachers have managed to overcome the barriers/obstacles generated by the transition from traditional to the online teaching system.

The proposed chapter is structured in five sections. The first section, the present one, highlights the preliminary aspects of the scientific approach. The second section highlights the background and the relevant scientific literature, and the next two sections present the research methodology, respectively a discussion on the results obtained. Finally, the fifth section draws the conclusions of our undertaken case study.

5.2 Background

The crisis we are going through today makes us pay more attention to the words of American molecular biologist Joshua Lederberg, Ph.D., a Nobel laureate who said that "the only threat to human domination on the planet is the virus." We noticed that all of a sudden, because of a virus, the world slowed down … businesses were closed, areas were quarantined and the life we considered normal was disrupted.

Due to measures taken in order to limit the spread of the novel coronavirus, teaching activities have been transposed into the online environment, under conditions that avoid close, in person meetings. Online learning is not a new concept, being a credible and effective form of learning [1], extremely popular in the last twenty years among the higher education sector who offers degree programs in online and distance modes of study [26]. Nevertheless, online education became even more widespread in pandemic times in which social distancing is imposed, the restrictive measures requiring a swiftly transition from traditional learning to online learning. Actually, e-learning represents a distance learning that combines classical learning methods with methods based on the use of technology, being the most accessible means for education, nowadays, under COVID-19 restrictions [17].

The changes made in the learning environment do not represent a matter of choice thus, the teaching activity had to be rethought so that the learning process would not be strongly disturbed. The effectiveness of distance e-learning during the COVID-19 outbreak is conditioned by a wide range of education-related factors that must be considered in this transition, namely the necessary adjustments of teaching methods and assessment techniques to the online environment, ICT access and usage and the individual's ability and motivation to study [19].

The measure imposed by the closure of schools in order to limit the spread of the novel coronavirus implied the transposition of the entire teaching activity in the online environment, respectively adapting the content of the course activities through more innovative teaching methods which required additional logistic assistance. The internet and emerging technologies made it possible to provide tutoring worldwide, creating a higher potential for the students to enroll in virtual classes [21].

Thus, videoconferences, audio and video recorded lectures or shared online materials have become effective tools in supporting the activity in the new context created by the COVID-19 pandemic [10]. However, this was not an easily process, not everyone was prepared for it, but represented the only way to solve the pedagogical catastrophe generated by the novel coronavirus outbreak [2].

In this context, e-Learning may become a part of the nowadays educational system, but it will not be sustainable unless those involved embrace it (Saha et al. 2020). On the one hand, studies state that online education consumes much more time, training and implication compared to traditional face to face classes [3]. Other research conducted by Fauzi and Khusuma [9] highlights that teachers understood the pandemic context but they felt dissatisfied teaching online. Also, the results highlight that online lectures were problematic too, because many issues were found, namely in the availability of facilities, internet usage or planning and learning evaluation. On the other hand, studies state that online teaching was preferred during the pandemic period by the vast majority of the teaching staff but regarding the post COVID-19 period only young university teachers will prefer to continue in the same manner because they are more familiar with modern technologies [24].

From the student's perspective, studies state that shifting to online learning caused technological, mental health, time management, and improper balance between life and education hours. Also, more than a third of the respondents declared that they were dissatisfied by the online experience [19].

Husain et al. [14] conducted another study in order to explore student's attitude towards online learning during COVID-19 pandemic and their research revealed both positive and negative aspects. The positive aspects highlighted increased participation at online courses, convenience, safety in uncertain times and effectiveness of time and costs. The negative aspects were represented by reduced focus and motivation, workload, technological problems and insufficient external support. [22] in the study conducted among Romanian students from "Vasile Alecsandri" University from Bacău highlighted, on the one hand, that most of students were satisfied with the measures taken during the lockdown period and the way the teaching—learning-assessment process took place. On the other hand, they reveal some negative effects of the online teaching experience like less effective teacher-student communication and interaction and the impossibility of performing practical applications. The same results were obtained by Ionescu et al. [15] which argue that students have accepted online learning, even if they find it less attractive than the traditional face-to-face education system.

In this context, we consider that the way in which the online learning system is perceived depends to a large extent on the interest given by teachers in adapting their materials and methods to the new digital school environment. Adequate space for study, internet access and technology are also important factors in the absence of which online learning could not be achieved. Also, from our point of view, the efficiency of online learning depends to a large extent on the student's interest, motivation and involvement.

With these nuances under consideration, this research study proposes the following main hypotheses:

H1: *the implications in teaching activities affects the perception of the opportunity of the e-learning system;*

H2: *IT and logistics support influences the perception of the opportunity of the e-learning system;*

H3: *the implications for learning activities affects the perception of the opportunity of the e-learning system*;

Those hypotheses are tested, limiting our research to Romanian academic environment, which among other countries, was not entirely prepared for transition to online education. This transition has been made in short time, reason why disruptions appeared from the beginning, concerning improper adjustment of didactical methods and tools, or sufficient knowledge on the area of emerging educational technologies and software solutions, or IT equipment availability [8, 15]. However, their results suggested that students seem to adapt in short time, in spite of the fact they see the online educational solution as less attractive compared with the traditional face-to-face educational system. Reasons behind are numerous. However, emphasis is made on the theory that poor students' perception on expected outcome of online educational systems, or awareness on psychological distress leads to lower attractiveness for online learning solutions [23].

On those circumstances, national authorities are invited to design and implement a clear mix of strategies and policies on the direction of educational resilience, which does not limit only to minimum requirements included on the recently approved PNRR for Romania, inviting all actors willing to bring their insights. Otherwise, Romanian educational system will just continue to deepen the inequalities on the educational system, as noted already in the mass-media and scientific reports made already public [13].

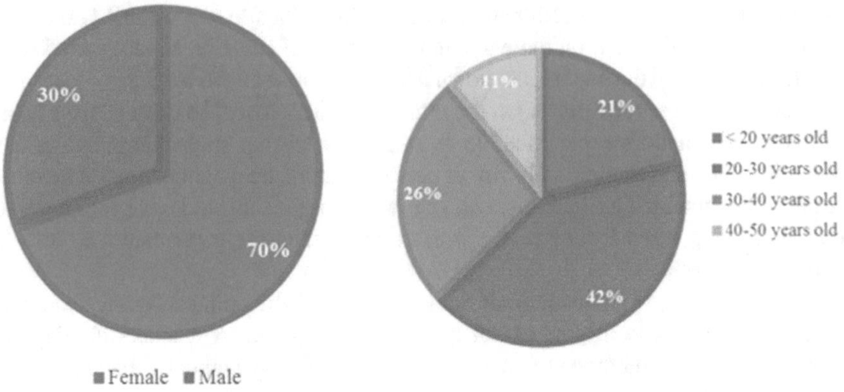

Fig. 5.1 Sample distribution by gender and age. *Source* Authors' projection

5.3 Methodology Research

5.3.1 Data Collection

The study starts from data collected through dissemination of a questionnaire among students following classes at West University of Timisoara, Faculty of Economics and Business Administration. Data collected measures students' perception on different implications of transition to online learning system. Items verified concern two levels of analysis, respectively students' perception concerning changes made on teaching models and students' perception about changes made on students learning models.

The questionnaire was submitted through google forms platform. On the December 2021–January 2022, we have received 89 questionnaires with no field left blank. In Fig. 5.1 we represented the sample composition of students sending back those questionnaires. On one hand, we see that the feedback was received mainly from female students (70%). On the other hand, sample distribution suggest that the major part of the responses were received from students aged under 30 years (63%), most of them aged between 20–30 years old (63%).

5.3.2 Reliability Analysis

The questionnaire submitted consist of 19 questions, synthesized in Table 5.1. The questions address both advantages (flexibility, adaptability, auto discipline etc.) and disadvantages (motivation, focus, software and educational technology, etc.) of online learning system.

Those items are included in the questionnaire, in order to measure students' perception concerning three main directions of research related to transformations

Table 5.1 Items short description

Item no	Item	Explanation	SMC[a]
1	Adaptability	Perception on how students' adaptability to changes	0.718
2	Auto discipline	Perception on how e-Learning improves auto discipline for students	0.770
3	Content	Perception on quality of content provided along the teaching process	0.658
4	Ease	How easy students can gather knowledge in the scenario of e-Learning	0.723
5	Equipment	Appreciation on the availability of proper equipment	0.472
6	Feedback	Rating on how students' perceive the feedback received from teachers	0.658
7	Flexibility	Perception on how students see e-Learning provide flexibility	0.707
8	Focus	Perception on how students' attention on learning activities is affected	0.619
9	Innovation	Are they used innovative teaching techniques in case of e-Learning	0.765
10	Interaction	Perception on how teachers interact and communicate with students	0.643
11	Internet	Students' perception on internet speed and technical related issues	0.469
12	Motivation	How students' motivation is affected by e-Learning	0.732
13	Place	Students rating on how suitable is the place for dedicated for eLearning	0.691
14	Results	Students' rating concerning impact of e-Learning on academic results	0.500
15	Practice	Students' perception on how seminar are limited	0.545
16	Storage	Students' perception on how e-Learning provide viable solution for storing learning materials	0.657
17	Technology	Perception on how students face with technological constraints (software knowledge, technology availability etc.) on e-Learning	0.594
18	Decision	Students' choice for e-Learning, hybrid or traditional system	0.430
19	Perception	Students' perception when comparing e-Learning other systems	0.694

[a] Squared Multiple Correlation
Source Authors' projection

determined by COVID-19 pandemic and transition to online learning on the academic environment, respectively students' perception on:

- changes determined on *teaching models* (content complexity and novelty, interaction with teachers during classes, use of innovative didactical methods and tools, practical component of seminar activities, classes flexibility);

- changes determined on *students' learning system efficiency* (ease of information learning, academic performance improvement, learning system adaptability) and some *psychological implications* (concentration, auto discipline, feedback from learning activity);
- *technological implications* (dedicated accounting and audit software, educational technologies and platforms, equipment availability, internet connection).

The questionnaire is designed considering Likert scale (1—"totally disagree", 2—"disagree", 3—"neutral", 4—"agree", 5—"totally agree"). For internal consistency of questionnaire design, we perform a scale reliability analysis. We obtain a Cronbach's Alpha statistic of 0.844, which is higher than the threshold of 0.80 [12], meaning our questionnaire is properly designed. Items are internally consistent, ensuring lower error in measuring respondents perception, no matter the sample considered [4]. Higher internal consistency suggests as well higher correlation between items and relevance for constructs validation [16].

5.3.3 Data Reduction

To reduce the volume of data collected through questionnaire dissemination, we have performed categorical principal components analysis on the two levels of analysis, respectively the impact of online learning on teaching models and students' learning systems. Categorical principal components analysis is proper to reduce nominal data, transforming it into numerical scale by using an estimated nonlinear non-monotonic function of transformation [20]. The method consists of optimal scaling of categorical data by assigning optimal scale values to each of the categories, in order that, for a specified number of dimensions, the overall variance accounted for the transformed variables is maximized.

Suppose we have n individuals for which we have collected scores for m questionnaire items. Data collected is represented into a S matrix where x_{ij} represent the score for individual i for item j. Those object scores are restricted by relation $S^T \cdot S = n \cdot I$, where I is the identity matrix. Object scores are centered values as well, as they are subject to restriction $1^T \cdot S = 0$, where 1^T is the vector of ones. Each nominal data x_{ij} is transformed into a quantified value, based on the scale dimensions identified according theoretical framework, using a function of transformationφ. Quantified scores $q_{ij} = \varphi(x_{ij})$ are standardized, considering the restriction $q_j^T \cdot q_j = n$. Those scores are multiplied by a set of optimal weights which are called component loadings. The matrix of component loadings A consist of m rows, similar with the number of items on the questionnaire, and p columns representing the number of components/dimensions identified. Those component loadings

$$L(Q, A, S) = \frac{1}{n} \cdot \sum_{j=1}^{m} trace\left(q_j \cdot a_j^T - S\right)^T \cdot \left(q_j \cdot a_j^T - S\right)$$

Maximization of variance accounted for the transformed scores consist in fact on the minimization of the loss function that measure the difference between original data and principal components, expressed by function above, using an alternative least squares algorithm [18].

5.3.4 Multinomial Regression Analysis

Further econometric analysis is performed to investigate the association between the dimensions identified running categorical principal components analysis (CATPCA). For this purpose, we estimate a multinomial logistic regression, as the dependent variable does not indicate any order between the different possible values [11]. The dependent variable is a nominal variable that can take three possible values, respectively classical learning system (value 0), online learning system (value 2), or hybrid learning system (value 3). Starting from the 3 possible outcomes, the procedure consists in fact on estimating two independent binary logistic regressions. In each of those models, we consider as reference the value corresponding to classical learning system, while the alternative is one of the alternative learning systems (online/hybrid).

In general, in case of k possible values for the dependent variable, we estimate $k - 1$ binary regression models expressed by relation below:

$$ln\frac{P(y_i = j)}{P(y_i = r)} = \beta_j \cdot x_i$$

where β_j is the vector of regression coefficients, $P(y_i = j)$ is the probability that outcome j is selected, whereas r is the "pivot" outcome.

Based on this odds ratio, we determine the probability that an individual changes his preference, from the r outcome to the new j preference, based on relation below:

$$P(y_i = j) = \frac{e^{\beta_j \cdot x_i}}{1 + \sum_{t=1}^{k-1} e^{\beta_j \cdot x_i}}$$

Additionally, we can determine the odds ratio determined by the ratio $\frac{P(y_i=j)}{1-P(y_i=j)}$ that show the chance that a respondent changes his initial option.

Further statistics evaluation is made, such as analysis on McFadden value, the Chi-Square test, or the percentage of correct classification made using estimated multinomial model [11]. Those statistics are relevant to analyzed the model fit and significance on explaining respondents' choice, based on the vector of input variables.

5.4 Results and Discussions

5.4.1 Student's Perception on COVID-19 Pandemic Implications for Learning Activities

The results collected from disseminating the questionnaire measuring student's perception on how COVID-19 pandemic has changes their way of learning and how teachers have reached to overcome the issues students' have faced on those pandemic times. In Table 5.2 we provide descriptive statistics on the distribution of responses students have submitted.

On one hand, the distribution shows that teachers have come with proper solutions during the pandemic period, that forced them to make transition of the teaching activities on online environment. Adapting the content of the course activities (88.76%), using more innovative teaching methods (65.17%), or keeping their interaction with the students (84.27%), during and outside teaching activities have been positively

Table 5.2 Items distribution of responses

Item	Totally disagree	Disagree	Neutral	Agree	Totally agree	Positive feedback (%)	Negative feedback (%)	Neutral feedback (%)
Perception	4	5	18	33	29	69.66	10.11	21.00
Content	1	1	8	21	58	*88.76*	2.25	9.00
Practice	17	8	29	24	11	39.33	28.09	*40.06*
Interaction		3	11	13	62	*84.27*	3.37	12.24
Innovation	11	5	15	32	26	65.17	17.98	19.07
Results	8	4	40	22	15	41.57	13.48	*49.13*
Ease	2	3	12	32	40	*80.90*	5.62	13.67
Adaptability	1	5	14	26	43	*77.53*	6.74	15.77
Feedback	17	9	29	18	16	38.20	29.21	*40.07*
Storage	22	1	13	27	26	59.55	25.84	19.23
Flexibility	21	1	7	30	30	67.42	24.72	10.19
Focus	24	9	17	21	18	43.82	*37.08*	25.98
Motivation	2	4	17	21	45	*74.16*	6.74	19.38
Auto discipline	15	4	20	28	22	56.18	21.35	26.82
Technology	21	17	26	17	8	28.09	*42.70*	*38.08*
Equipment	1	3	4	11	70	*91.01*	4.49	4.50
Internet speed	13	2	12	25	37	69.66	16.85	15.65
Place		3	14	27	45	*80.90*	3.37	15.59

Source Authors' calculation

perceived by most of the participants to the questionnaire, similar results with the ones obtained by Radu et al. [22] and Ionescu et al. [15].

On the other hand, students did not yet made a clear opinion on the limitations to the output expected by students from seminar activities during online teaching (40.06%). However, this results involves a more in-depth discussion, as root-causes are multiple, such as insufficient connection between seminar activities and practical requirements claimed by companies, deterioration of students' attention essential on getting a proper understanding of seminars content, or maybe the insufficient digitalization of accounting, financial reporting and auditing processes.

Similar positive results are suggested concerning learning activities as well, as most of the items included in the questionnaire disseminated are positively rated by respondents, especially in terms of new information assimilation (77.53%), adaptability and motivation (74.16%). Instead, slight portion of respondents claim that activities performed online affects negatively their focus (37.08%). Additionally, they emphasize that for authorities should think more strategically and support efforts on providing guidance on new technology used on teaching activities, as the barriers generated by the lack of little knowledge about different solutions of education software, or educational technologies may deter the outcome of the teaching activities, with implications on students' learning activities effectiveness.

However, seems that students started more and more to embrace the solution of online learnings, including the alternative e-Learning products and services, as only 10.11% of the respondents have considered online system is yet not properly designed and implemented to be considered for further use.

Overall, in Fig. 5.2 we analyze the correspondence between the most frequent rating per items included in the questionnaire disseminated and students' rating. In Table 5.3 we provide basic statistics for the correspondence analysis performed and represented on Fig. 5.2, showing that results are statistically significant ($\aleph^2 = 583.91, Sig. < 0.01$). The two dimensions reflected on the plot below explain approximately 86.7% of the variation on students' responses, which is shows the reduction of dimension on our analysis is valid and does not alter the interpretation of the results.

We observe that the highest rating translated into extremely positive perception concerning viability of online learning system reduces to the items reflecting the content of teaching activities (item 3), or the way teachers have overcome the obstacles on communication with the students during COVID-19 pandemic (item 10).

Instead, most of the items are rated positively (e.g.: item 7, item 9), but with some reserves (rating 4), which suggest that students ask for continuous improvement and systemic solutions to the current challenges and opportunities on designing and implementing online learning systems. Moreover, students realize that they have to address as well the challenges and opportunities related to how they can improve their adaptability to changes (item 1), auto discipline (item 2), or accumulation of new academic content in alternative learning systems (item 4). For this purpose, they most probably should start from increasing their motivation when following online

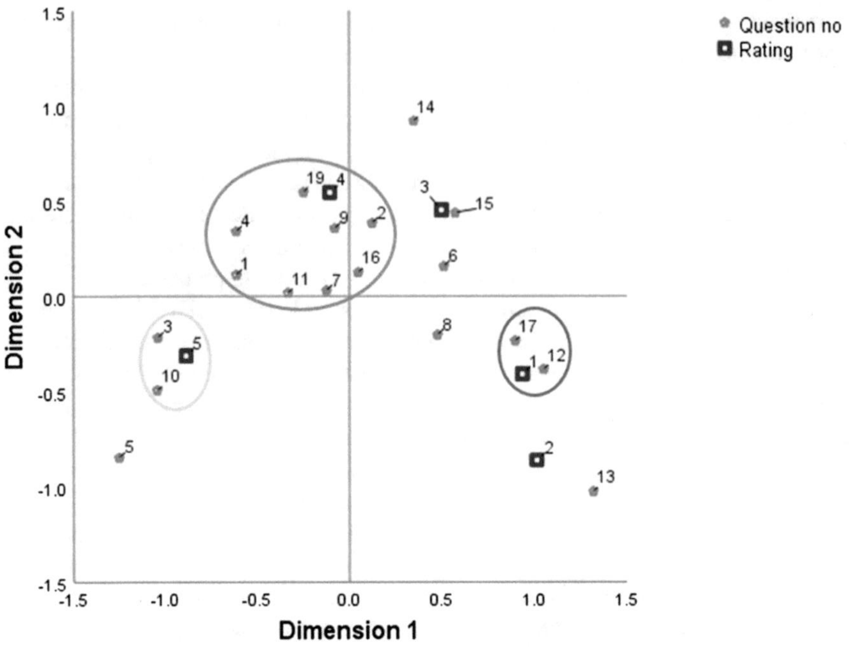

Fig. 5.2 Correspondence analysis on students' rating distribution on addressed items. *Source* authors' projection with SPSS 19.0

Table 5.3 Correspondence analysis statistics

Dimension	Singular value	Inertia	Proportion of inertia	
			%	Cumul. %
1	0.514	0.264	0.726	0.726
2	0.227	0.052	0.142	0.867
3	0.192	0.037	0.101	0.969
4	0.107	0.011	0.031	1.000
Total		0.364	1.000	1.000
Chi square	583.908			
Sig.	0.000[a]			

Source Authors' calculation
[a] 72 degrees of freedom

courses and seminars as seem this is an essential factor from students' perception that might affect learning outcome (item 17).

Nonetheless, we see that students face less issues with the internet connection (item 11), or learning resource materials storage (item 16). Instead, the underline that online learning solutions raise them serious limitations on gathering new academic insights, because of upcoming educational technologies and software (item 17).

Those education technologies are more related to solutions specific for the area of the activity, rather than general platforms used for online teaching activities.

We continue the statistical exploratory analysis with descriptive statistics on the items included on the questionnaire, provided in Table 5.4. The results a relatively positive perception students have about the viability of online learning system, as the mean of 3.876 is closer to rating 4 from our Likert scale, reflecting agreement with reserves. Students seem to have a relatively similar perception on this direction, as the standard deviation of only 1.075 is not high when using Likert scale with 5 levels. Instead, when asked what learning system they would choose for the future, the opinions split drastically, compared with students' perception, as the mean of 0.989 and the higher standard deviation of 0.612 show high level of heterogeneity.

In Fig. 5.3 we observe that the issue is students' do not unanimously express their choice for online learning system. They actually perceive the hybrid learning system a real alternative solution for the future, especially when looking to the students that have expressed a neutral rating (38.89%).

Heterogeneity on students' responses is visible also in case of item addressing the limitations that online learning systems might generated on the practical components

Table 5.4 Items descriptive statistics

	Item	Mean	Percentiles		Std. dev.	Coef. variation
			1st	3rd		
	Decision	0.989	1.00	1.00	0.612	0.619
	Perception	3.876	3.00	5.00	1.075	0.277
Teaching process	Content	4.506	4.00	5.00	0.799	0.177
	Practice	3.045	2.00	4.00	1.278	***0.420***
	Interaction	4.506	4.00	5.00	0.841	0.187
	Innovation	3.640	3.00	5.00	1.299	0.357
Learning process	Results	3.360	3.00	4.00	1.100	0.327
	Ease	4.180	4.00	5.00	0.948	0.227
	Adaptability	4.180	4.00	5.00	0.972	0.233
	Feedback	3.079	2.00	4.00	1.342	0.436
	Storage	3.382	1.50	5.00	1.534	***0.454***
	Flexibility	3.528	2.50	5.00	1.545	***0.438***
	Focus	3.000	1.00	4.000	1.500	***0.500***
	Motivation	4.157	3.00	5.00	1.032	0.248
	Auto discipline	3.427	3.00	4.50	1.364	0.398
e-Learning logistics	Technology	2.708	2.00	4.00	1.272	0.470
	Equipment	4.640	5.00	5.00	0.815	0.176
	Internet speed	3.798	3.00	5.00	1.391	0.366
	Place	4.281	4.00	5.00	0.853	0.199

Source Authors' calculation

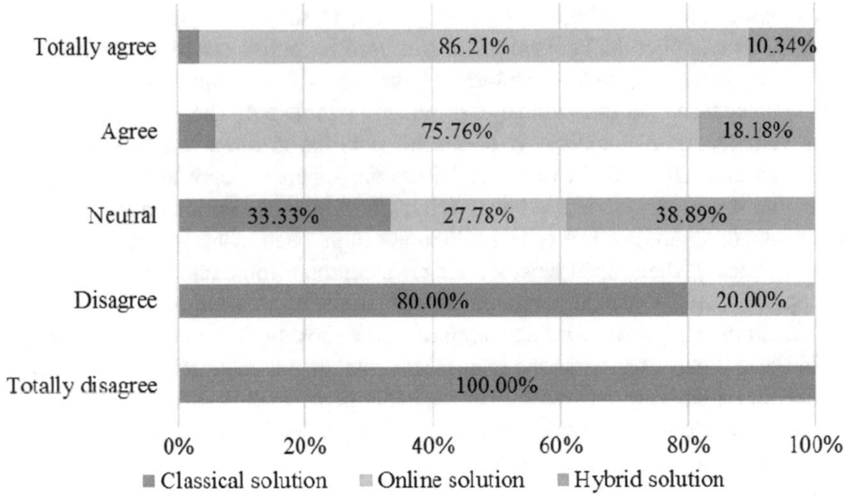

Fig. 5.3 Correspondence analysis on students' rating distribution on items addressed. *Source* Authors' projection

of academic programs ($Coef.var. = 0.420$), perception on flexibility ($Coef.var. = 0.438$), lack of attention ($Coef.var. = 0.500$), or technological issues and challenges ($Coef.var. = 0.470$).

If the issues suggested on the area of technology address the same root-cause for most of the students, the other items show in fact the implications of different learning strategies and techniques students use, which are subject to students' action and less educational system which might only recommend such strategies, techniques and even tools.

5.4.2 *Premises and Connections Relevant on Students' Perception Concerning Online Learning*

Students' perception on online learning solution viability and option for one of the three versions of learning systems, represents subjective measures highly influenced by a cumulus of factors. Those factors address the problem of teaching model, highly dependent on teachers' actions, and learning strategies as well, conditioned by student' abilities and aptitudes. In Table 5.5 we provide results on Spearman correlation, to have a better understanding on the association between items addressing different criteria students look for when making an opinion.

Results show that association with students' perception viability of online learning system is higher only in case of adaptability item (0.582). This association shows that students have a positive perception on the online system especially when they believe they can adapt timely. Otherwise, they fear about the negative effects on the academic

Table 5.5 Correlation matrix

	Adaptability	Decision	Flexibility	Innovation	Interaction	Perception	Technology
Decision	0.223*	1.000	0.083	0.219*	0.160	0.268*	0.022
Perception	***0.582***	0.268*	0.254*	0.315**	0.451**	1.000	−0.108
Content	0.277**	0.063	0.143	0.403**	***0.757***	0.451**	0.013
Practice	−0.104	0.039	0.285**	0.270*	−0.132	−0.084	***0.557***
Interaction	0.324**	0.160	0.110	0.328**	1.000	0.451**	−0.028
Innovation	0.293**	0.219*	***0.710***	1.000	0.328**	0.315**	0.292**
Results	0.387**	0.120	0.011	0.063	0.199	0.421**	−0.235*
Ease	***0.778***	0.088	0.202	0.272**	0.352**	***0.554***	−0.094
Adaptability	1.000	0.223*	0.236*	0.293**	0.324**	***0.582***	−0.146
Feedback	0.377**	0.239*	0.477**	***0.660***	0.313**	0.400**	0.067
Storage	0.307**	0.246*	***0.686***	***0.681***	0.161	0.336**	0.095
Flexibility	0.236*	0.083	1.000	***0.710***	0.110	0.254*	0.192
Focus	−0.455**	−0.250*	−0.177	−0.282**	−0.429**	***−0.673***	0.113
Motivation	0.490**	0.082	0.072	0.247*	***0.605***	***0.688***	−0.094
Auto discipline	0.323**	0.244*	***0.661***	***0.743***	0.232*	0.316**	0.025
Technology	−0.146	0.022	0.192	0.292**	−0.028	−0.108	1.000
Equipment	0.245*	0.127	0.068	0.214*	0.406**	0.314**	−0.028
Internet speed	0.203	0.116	***0.658***	***0.549***	0.169	0.201	0.170
Place	0.172	0.191	***0.677***	***0.581***	0.113	0.294**	0.290**

Source Authors' calculation

results, as the correlation of 0.451 suggest that students are more willing to choose in favor on online learning system as long as they get the support from teachers on understanding the new information during classes and even through online platforms that ensure open channels for permanent communication with teachers. This communication and interaction with the teachers is reflected on the way students' perceive the content presented during classes (0.757). On this direction, a solution already suggested by students is that innovative didactical methods and tools to be considered by teachers during classes, which could improve flexibility of the classes (0.710), or provide better understanding for the students through transmitting continuous feedback to students (0.660). Hence, students' motivation can increase knowing there is somebody guiding them all the time (0.605).

5.4.3 Dimensions Considered on Students' Perception Configuration

In Table 5.6 we describe the basic statistics for the principal components analysis performed to reduce the data collected through the questionnaire to only few dimensions of analysis. We have decided to reduce our analysis to a two-dimension analysis when assessing the association between students' option for one of the three learning systems included on the questionnaire and their perception on the items describing their perception on how teaching activities and learning activities have changed together with transition to online learning.

The results show that the two dimensions obtained per each group of items from the questionnaire exceed to cover the variation on students' responses the percentage of 60%. On one hand, dimensions describing the items reflecting students' perception on changes determined on teaching model, approximately 62.88% of the variation in our sample is covered. On the other hand, dimensions describing the items reflecting students' perception on changes determined on their learning models, approximately 67.63% of the variation in our sample is covered.

In Fig. 5.4 we represent the items included on the two models of principal components analysis. On the right side is provided the plot for the PCA concerning changes on teaching models, while the left plot describes the PCA reducing the items related to changes on learning models.

In case of PCA related to students' learning models the second dimension describes better the disadvantages of online learning, such as the negative effects on students' motivation for learning and the concentration needed for learning activities, especially when they do not have available a proper place for participating to online classes or performing the learning activities assigned.

The first dimension resulted from this PCA comprise more items, but rather reflects information on students' perception about the ease of assimilating information on the learning activities, the level of academic results obtained through online learning, or about students' capacity to adapt to changes involved by transition to online learning.

Table 5.6 Categorical principal components analysis statistics

	Dimension	Cronbach's alpha	Variance accounted for	
			Total (Eigenvalue)	% of variance
Teaching process	1	0.762	2.999	37.493
	2	0.580	2.031	25.384
	Total	0.916[a]	5.030	***62.877***
Learning process	1	0.868	4.575	45.751
	2	0.603	2.188	21.882
	Total	0.947[a]	6.763	***67.633***

[a] Total Cronbach's Alpha is based on the total Eigenvalue
Source Authors' calculation

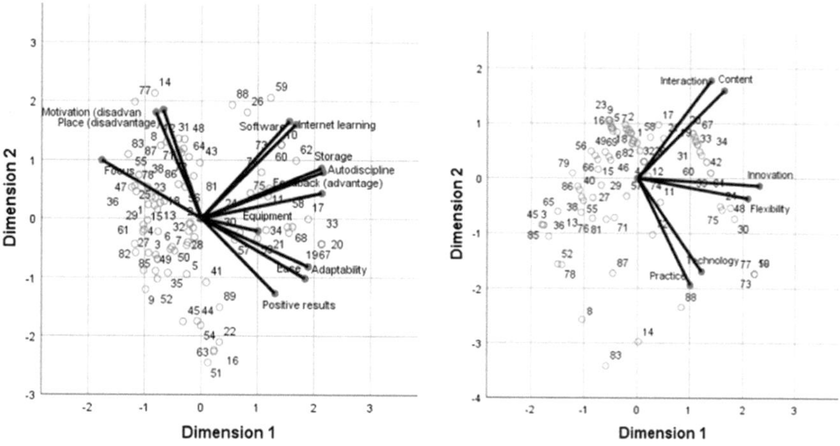

Fig. 5.4 Representation of PCA dimensions and items reduced. *Source* Authors' calculation

Interesting results are related to technological constraints implied by this transition to online learning, as both items concerning software and internet speed seem to be of same importance for both dimensions. This result conforms that challenges and opportunities on professional software solutions and emerging technologies affect both the students' willing to learn and adapt to changes. We believe that this is one root-cause for the results that show significant part of the students disseminating our questionnaire prefer a hybrid learning system. They realize the benefits generated by the online learning system, but they are aware as well about the insufficient guidance on emerging technologies and dedicated software solutions that allow online learning activities to determine expected outcome.

In case of PCA related to students' perception on how teaching models change, the second dimension is more related to items concerning students' perception on characteristics of teaching model, as it affects students' ease of assimilating new academic information, such as the complexity of academic content submitted, or the way it is presented and interaction with students for proper understanding is ensured.

Instead the first dimension is more related to the practical component of teaching activities and emerging technologies used for illustration during classes. Higher applicability of theoretical information is expected to be illustrated through case studies supported by dedicated software solutions and emerging technologies supporting educational activities.

Therefore, in case of PCA results, we rename our factors suggestively as follows:

- in case of PCA on items related to students' perception on changes made by transition to online learning system, on the effectiveness of learning activities:
  - first dimension is renamed as *learning motivation*, those items being strongly correlated, as the level of students' motivation is a premises for higher concentration (focus) of learning activities;

  - second dimension is renamed *learning effectiveness*, as the item of ease reflect actually the outcome of students' learning activities; the item of adaptability is conditioned as well by the same outcome of learning activities, as through learning activities students should reach to an independent and critical thinking that allows them to adapt easier to any further changes;

- in case of PCA on items related to students' perception on changes made by transition to online learning system, on teaching activities that affect students' effectiveness of learning activities:

  - first dimension is renamed *technological constraints*, as it incorporates the symbiosis between emerging technologies supporting educational activities and the requirements of practical approach of classes in the perspective of higher digitalization of firms' processes;

- second dimension is renamed as *teaching effectiveness*, as the scope of the teaching activities is that students' assimilate the content presented during classes, using proper didactical methods, techniques and tools to ensure strong interaction with the students.

5.4.4 Drivers Analysis from a Dimensional Simplified Approach

Based on the four dimensions determined running categorical PCA for both levels of analysis, respectively changes on teaching models and changes on learning activities, we present in Table 5.7 results of the multinomial logistic regression. This way, we perform an econometric analysis on the marginal effect of each of those dimensions on odds that students' change their option for one learning system.

The results show that marginal statistically significant effect on the odds that students' change their option in favor of classical learning system is determined only by students' perception about the online learning system and the PCA dimension describing the measure of teaching effectiveness on students' perception based on experimenting online courses and seminars.

Students' perception on how relevant are for learning activities the online courses and seminar determine a positive marginal effect on the odd that a student would change his option for an online learning system ($Coef. = 2.041, Sig. < 0.05$), compared with the option for the classical face-to-face learning system. Therefore, in case students perceive the online teaching activities improve their capacity to learn, the odds to change their option in favor of online learning system increase more than seven times ($e^{2.041} \cong 7.698$).

Similar results are obtained in case of the odds students have in favor of hybrid learning system, but with a slightly lower marginal effect ($Coef. = 1.285, Sig. < 0.10$), meaning an increase of the odds with approximately three times ($e^{1.285} \cong 3.614$) as well. The slight difference of marginal effect of learning effectiveness on

Table 5.7 Categorical principal components analysis statistics

Dependent variable: decision[a]		B	Std. error	Wald	df	Sig.	Exp (B)
Online learning system	Intercept	−6.869	2.915	5.554	1	0.018	
	[Gender = 0]	0.083	1.012	0.007	1	0.935	1.086
	[Gender = 1]	0[b]			0		
	[Age = 15]	0.983	1.596	0.379	1	0.538	2.672
	[Age = 25]	1.409	1.466	0.924	1	0.337	4.092
	[Age = 35]	0.419	1.463	0.082	1	0.774	1.521
	[Age = 45]	0[b]			0		
	Perception	*2.041*	*0.701*	*8.485*	*1*	*0.004*	*7.698*
	Technological constraints	−0.019	0.591	0.001	1	0.975	0.981
	Teaching effectiveness	*1.203*	*0.569*	*4.477*	*1*	*0.034*	*3.332*
	Learning effectiveness	0.077	0.817	0.009	1	0.925	1.080
	Learning motivation	−0.200	0.683	0.086	1	0.770	0.819
Hybrid learning system	Intercept	−2.492	2.379	1.097	1	0.295	
	[Gender = 0]	−0.651	1.004	0.420	1	0.517	0.522
	[Gender = 1]	0[b]			0		
	[Age = 15]	−2.143	1.410	2.310	1	0.129	0.117
	[Age = 25]	−1.763	1.445	1.490	1	0.222	0.171
	[Age = 35]	−1.450	1.449	1.001	1	0.317	0.235
	[Age = 45]	0[b]			0		
	Perception	*1.285*	*0.685*	*3.516*	*1*	*0.061*	*3.614*
	Technological constraints	−0.740	0.608	1.480	1	0.224	0.477
	Teaching effectiveness	0.382	0.553	0.477	1	0.490	1.465
	Learning effectiveness	0.626	0.850	0.541	1	0.462	1.869
	Learning motivation	1.155	0.748	2.382	1	0.123	3.174

[a] The reference category is: 0
[b] This parameter is set to zero because it is redundant
Source Authors' calculation

the odds students might choose hybrid learning systems instead of the traditional learning systems, compared with the odds students would prefer online learning system, can be associated to the fact the interaction with teachers in case of hybrid system improves because of face-to-face meetings with students.

The effectiveness of online classes based on students' perception seem to affect significantly only the odds that students would change their preference from classical learning system to online learning system ($Coef. = 1.203$, $Sig. < 0.05$). This results confirms indirectly how important is the interaction between students and teachers, especially when the content presented during classes is of higher complexity

Table 5.8 Categorical principal components analysis statistics

Statistic	Value
Cox and Snell	0.527
Nagelkerke	0.628
McFadden	0.409
Chi-square	66.68
Df	18
Sig.	0
Classification percentage	0.798

Source Authors' calculation

or novelty, including academic insights on the area of merging technologies of accounting and audit software solutions that support firms' processes digitalization, such as big data analytics, system audits, internal controls digitalization etc.

Instead, seems that students' choice for an alternative learning system does not depend on how they design their learning strategies, which suggest us students do not adjust drastically their learning activities when transition to online learning is made. Moreover, negative effects of online classes on students' motivation does not impact significantly their preference for an alternative learning system, results which suggest us students have set-up sufficiently clear learning objectives that provide them motivation to succeed, no matter the support offered by teachers during online classes. On those circumstances, we appreciate that online courses improve on a wider perspective students' auto discipline, making them more responsible and aware of the role of the educational activities, either we talk about independent learning activities, or learning activities supported by university online courses.

In Table 5.8 we provide basic statistics for the multinomial regression model estimated in our study. The McFadden value of 0.409 seem to be sufficiently high [11], suggesting that the model is representative on reflecting the association between students' preference for an alternative learning system and their perception on different implications determined by transition to online learning system. Moreover, the Chi-Square test is statistically significant, which show that the model is significantly different from the multinomial logistic regression model incorporating only an intercept. Nonetheless, we observe a good classification ratio, ensured by the model estimated, as 79.8% of cases are correctly classified [31]. Overall, we appreciate the model is statistically significant.

5.5 Conclusions

The pandemic generated by the new coronavirus has brought important changes in terms of teaching activities. By transposing the courses in the online environment, both teachers and students had to adapt and create a favorable learning environment.

Our research starts from data collected through dissemination of a questionnaire among students following classes at West University of Timisoara, Faculty of Economics and Business Administration. Data collected measures students' perception on different implications of transition to online learning system. Items verified concern two levels of analysis, respectively students' perception concerning changes made on teaching models and students' perception about changes made on students learning models.

The results show that teachers have come with proper solutions during the pandemic period, that forced them to make transition of the teaching activities on online environment. Adapting the content of the course activities (88.76%), using more innovative teaching methods (65.17%), or keeping their interaction with the students (84.27%), during and outside teaching activities are have been positively perceived by most of the participants to the questionnaire, being in line with the results obtained by Radu et al. [22] and Ionescu et al. [15].

Finally, we conclude that students face less issues with the internet connection or learning resource materials storage. Instead, the underline that online learning solutions raise them serious limitations on gathering new academic insights, because of upcoming educational technologies and software. Those education technologies are more related to solutions specific for the area of the activity, rather than general platforms used for online teaching activities.

References

1. Baciu, D., Ardelean, B.O., Ivascu, L., Fodorean, D.: Educația prin E-Learning, Ed. Academiei Oamenilor de Știință din România, Ed. Tehnică (2020). ISBN: 978-606-8636-73-3
2. Basilaia, G.: Replacing the classic learning form at universities as an immediate response to the COVID-19 virus infection in Georgia. Int. J. Res. Appl. Sci. Eng. Technol. **8**(3), 101–108 (2020)
3. Bussmann, S., Johnson, S.R., Oliver, R., Forsythe, K., Grandjean, M., Lebsock, M., Luster, T.: On the recognition of quality online course design in promotion and tenure: a survey of higher ed institutions in the western United States. Online J. Dist. Learn. Adm. **20**(1) (2017)
4. Catoiu, I., Balan, C., Popescu, I.C., Orzan, G., Veghes, C., Danetiu, T., Vranceanu, D.: Marketing research, Uranus (2002)
5. CDC: Centers for Disease Control and Prevention, Coronavirus Disease (2020). https://www.cdc.gov/
6. Cosmulese, C.G., Grosu, V., Hlaciuc, E., Zhavoronok, A.: The influences of the digital revolution on the educational system of the EU countries. Mark. Manag. Innov. **3**, 242–254 (2019)
7. Dictionary of Epidemiology. Oxford University Press (2008)
8. Edelhauser, E., Lupu-Dima, L.: Is Romania prepared for eLearning during the COVID-19 pandemic? Sustainability **12**, 5438 (2020)
9. Fauzi, I., Sastra Khusuma, I.H.: Teachers' elementary school in online learning of COVID-19 pandemic conditions. Jurnal Iqra': Kajian Ilmu Pendidikan **5**(1), 58–70 (2020)
10. Favale, T., Soro, F., Trevisan, M., Drago, I., Mellia, M.: Campus traffic and eLearning during COVID-19 pandemic. Comput. Netw. **176**(May) (2020)
11. Garson, G.D.: Logistic Regression: Binary & Multinomial. Statistical Associates Publishing (2014)

12. Hair, J.F., Jr., Black, W.C., Babin, B.J., Anderson, R.E.: Multivariate Data Analysis, 8th edn. Pearson (2019)
13. Hosszu, A., Rughinis, C.: Digital divides in education. An analysis of the Romanian public discourse on distance and online education during the COVID-19 pandemic. Sociologie Romaneasca **18**(2), 11–39 (2020)
14. Husain, B., Idi, Y.N., Basri, M.: Teachers' perceptions on adopting e-learning during Covid-19 outbreaks; advantages, disadvantages, suggestions. Jurnal Tarbiyah **27**(2) (2021)
15. Ionescu, C.A., Paschia, L., Nicolau, N.L.G., Stanescu, S.G., Stanescu, V.M.N., Coman, M.D., Uzlau, M.C.: Sustainability analysis of the E-learning education system during pandemic period—COVID-19 in Romania. Sustainability **12**, 9030 (2020)
16. Labar, A.V.: SPSS for Education Science. Polirom (2008)
17. Kaup, S., Jain, R., Shivalli, S., Pandey, S., Kaup, S.: Sustaining academics during COVID-19 pandemic: the role of online teaching-learning. Indian J. Ophthalmol. **68**, 1220–1221 (2020)
18. Linting, M., van der Kooij, A.: Nonlinear principal component analysis With CATPCA: a tutorial. J. Pers. Assess. **94**(1), 12–25 (2012)
19. Maqableh, M., Alia, M.: Evaluation online learning of undergraduate students under lockdown amidst COVID-19 pandemic: the online learning experience and students. Child. Youth Serv. Rev. **128** (2021). ISSN 0190-7409
20. Meulman, J.J., van Der Kooij, A.J., Heiser, W.J.: Principal Components Analysis with Nonlinear Optimal Scaling Transformations for Ordinal and Nominal Data, Part of The SAGE Handbook of Quantitative Methodology for the Social Sciences (2004)
21. Murday, K., Ushida, E., Chenoweth, N.A.: Learners' and teachers' perspectives on language online. Comput. Assist. Lang. Learn. **21**(2), 125–142 (2008)
22. Radu, M.C., Schnakovszky, C., Herghelegiu, E., Ciubotariu, V.A., Cristea, I.: The impact of the COVID-19 pandemic on the quality of educational process: a student survey. Int. J. Environ. Res. Public Health 23, **17**(21), 7770 (2020)
23. Roman, M., Plopeanu, A.P.: The effectiveness of the emergency eLearning during COVID-19 pandemic. The case of higher education in economics in Romania. Int. Rev. Econ. Educ. **37** (2021)
24. Saha, S.M., Pranty, S.A., Rana, J., Islam, J., Hossain, M.: Teaching during a pandemic: do university teachers prefer online teaching? Heliyon **8**(1) (2022)
25. Schleicher, A.: The impact of COVID-19 on education insights from education at a glance (2020)
26. Stone, C.: Online learning in Australian higher education: opportunities, challenges and transformations. Stud. Success **10**(2), 1–11 (2019)
27. United Nations: A UN framework for the immediate socio-economic response to COVID-19 (2020). https://unsdg.un.org/resources/un-framework-immediate-socio-economic-response-covid-19
28. UN: Policy brief: education during COVID-19 and beyond (2020)
29. UNESCO: COVID-19 educational disruption and response (2020). https://en.unesco.org/covid19/educationresponse. Accessed 14 Dec 2020
30. UNESCO: When schools shut: gendered impacts of COVID-19 school closures (2021). https://unesdoc.unesco.org/ark:/48223/pf0000379270. Accessed 14 Dec 2020
31. Verbeek, M.: A Guide to Modern Econometrics, 5th ed. Wiley (2017)
32. WHO: World Health Organization (2020). https://www.who.int/

Chapter 6
Integrated System for New Product Development in Healthcare

Andreea Ionica and Monica Leba

Abstract The chapter presents an integrated system concept, starting from the need for a rigorous approach regarding new healthcare product development. The chapter contains a theoretical part about machine learning in healthcare, Agile approach for healthcare organizations' projects and model based on the modified QFD method for Agile applied for healthcare products. The practical part reveals the application of the integrated system, including two case studies for innovative new healthcare products development.

Keywords New product development · Agile · SCRUM · QFD · Machine learning · Offset · Healthcare · Exoskeleton · Burnout device

6.1 Introduction

We live in a world of technology, which indicates the pace of development of new products. Thus, emerges the need to move towards agile approaches that respond quickly to the various needs, expressed or not, and the integration of modern machine learning technologies is essential for obtaining competitive products.

In the development of new products, the analysis stage is vital to obtain the most suitable product based on the requirements or needs of potential users. The design of a machine learning model responds very well to this stage of analysis, by creating a model that learns from existing data and based on them provides the best possible predictions.

A. Ionica (✉)
Department of Management and Industrial Engineering, University of Petrosani, Universitatii, 20, 332006 Petrosani, Romania
e-mail: andreeaionica@upet.ro

M. Leba
Department of System Control and Computer Engineering, University of Petrosani, Petrosani, Romania
e-mail: monicaleba@upet.ro

L. Ivascu et al. (eds.), *Intelligent Techniques for Efficient Use of Valuable Resources*, Intelligent Systems Reference Library 227,
https://doi.org/10.1007/978-3-031-09928-1_6

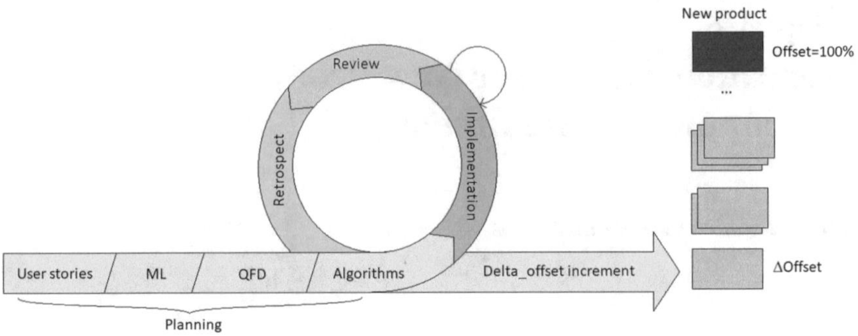

Fig. 6.1 New product development approach

The chapter proposes an integrated system for the development of new healthcare products with the following components (Fig. 6.1):

- SCRUM methodology from Agile approach;
- Model based on the modified Quality Function Deployment (QFD) method for Agile Software Development Life Cycle (SDLC);
- Optimal tasks planning algorithm;
- New Product Development (NPD) based on machine learning.

The specificity of the system for the healthcare field is given by the machine learning algorithms, therefore the structure of the chapter will contain a presentation of machine learning in healthcare together with the presentation of the elements of the integrated system and new developed products.

6.2 Machine Learning in Healthcare

6.2.1 Generalities

The first variants of using machine learning in healthcare were of the type:

- applications for care delivery improvement, such as drug time alarm applications,
- applications for clinical data management,
- real-time clinical monitoring and analysis devices capable of transmitting alarm signals when abnormalities are detected,
- medical devices for the early detection of health problems,
- medical devices for assessing the risks related to chronic diseases.

Since the advent of modern artificial intelligence and machine learning, mathematical algorithms have been developed to make predictions for a wide variety of fields of applications, including healthcare, such as decision-making, the development of new drugs, and image analysis, like the case of diabetic retinopathy.

Although it has been intuited since the 1970s that developments in the field of applied computers in medicine will increase and even replace the decisions of clinicians, to use machine learning to its full potential has had to wait for technological advances in recent years, such as the development of GPUs (Graphical Processing Units) which ensures parallel processing of large volumes of data at very high speeds and the availability of large volumes of digital healthcare data.

In addition to HOW, WHY is of the same importance in explaining machine learning's involvement in medicine.

Healthcare systems try to provide the best quality services to as many beneficiaries as possible in all geographical areas. At the same time, efforts are being made to reduce resources, costs and adverse effects on healthcare providers. All these answer to WHY. Machine learning provides support for achieving the above goals. Another motivation would be the fact that we are witnessing an accentuated aging of the population that is expected to continue in the coming years, which implicitly leads to a crisis of healthcare specialists. In addition, there are billions of people in the world who do not have access to basic healthcare services, which will motivate the development of artificial intelligence telemedicine services. The development of machine learning-based applications and devices will never replace the clinician, but will increase medical expertise for healthcare services to a larger population with a limited set of resources.

Machine learning applications in healthcare are virtually unlimited. These can be for:

- screening and diagnosis
- adaptive clinical trials
- global health
- home health and wearables
- drug discovery and design
- precision medicine
- robotics.

6.2.2 General Steps

In order to develop a machine learning based healthcare product, the main steps to be followed are:

A. Dataset dividing into: Training dataset (70%), Validation dataset (15%), Testing dataset (15%)

B. Machine learning model selection

In order to choose the best fitted machine learning model, in Fig. 6.2 is presented in a synthetic manner the main models applicable in healthcare classified in two categories, four models for prediction and four models for extracting insights. Also, there are presented examples of applications for each type of model.

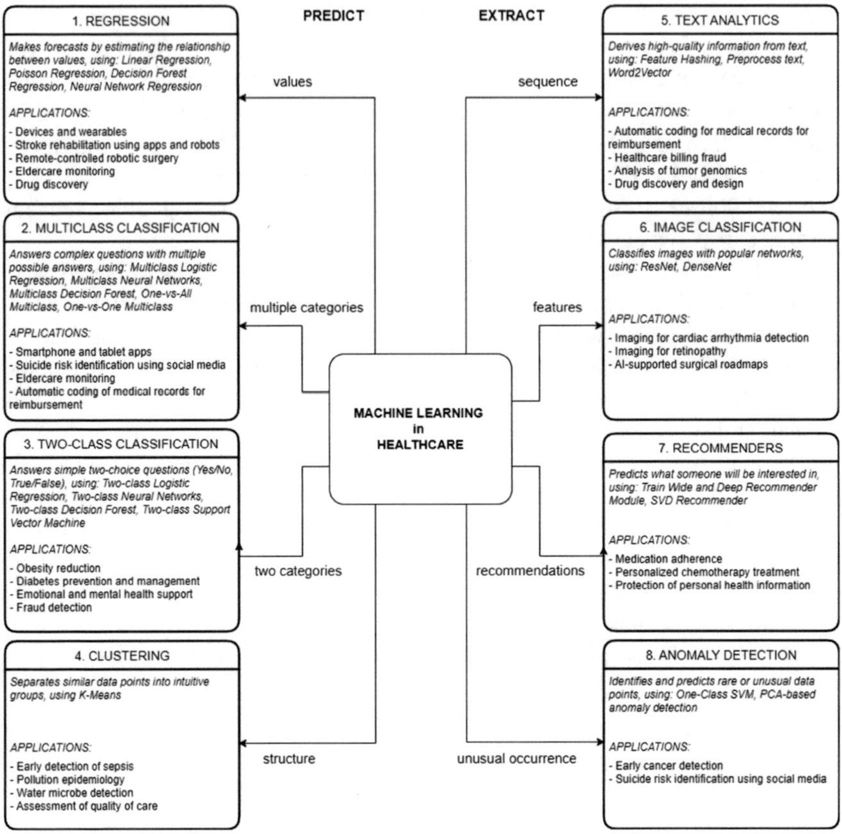

Fig. 6.2 Machine learning models in healthcare

In Fig. 6.3 there are resumed the above machine learning models and applications in healthcare with an emphasis on the actors involved in product development user stories acquisition.

C. Running on the training dataset

The chosen machine learning model is applied on the training dataset.

D. Calculation of the loss

It is an important step in optimizing the model, it is the one that decides when the model has completed the learning stage and the prediction errors are within an acceptable range. The goal of model training is to find the best set of parameters that leads to the least error for all examples in the dataset.

The mechanism of use of the loss is the evaluation of the differences between the predictions made by the model based on the data and the correct labels and based on these differences making the adjustment of the parameters to minimize the loss. The loss function is a mathematical function that processes

	Device product developers, clinicians and users	Clinician care teams	Public health program managers	Healthcare administrators	Geneticist	Pharmacologist
1. Regression	Health monitoring Rehabilitation	Surgical procedures	Population health			Drug discovery
2. Multi-class Classification	Benefit/risk assessment		Identification of individuals at risk Population health	International classification of diseases		
3. Two-class Classification	Disease prevention and management			Fraud detection		
4. Clustering		Patient safety	Population health	Physicians' management		
5. Text analytics				International classification of diseases Fraud detection	Genomics	Drug discovery
6. Image Classification		Early detection, prediction and diagnostic tools Surgical procedures				
7.Recommenders	Medication management	Precision medicine		Cybersecurity		
8. Anomaly Detection		Early detection, prediction and diagnostic tools Patient safety	Identification of individuals at risk			

Fig. 6.3 Actors' categories for user stories acquisition

all individual losses and provides a picture of the performance of the entire model.

Types of loss functions:

- Mean Squared Error Loss (MSE) is the simplest and most common and suitable especially for linear regression models. Squaring is to eliminate negative differences and to make sure that there are no outlier predictions that would generate large errors. The formula is:

$$MSE = \frac{1}{n} \cdot \sum_{1}^{n} (model\ prediction - ground\ truth)^2$$

- Mean Squared Logarithmic Error Loss

Mean Absolute Error Loss (MAE). It is similar to MSE without giving so much importance to outliers. The formula is:

$$MSA = \frac{1}{n} \cdot \sum_{1}^{n} |model\ prediction - ground\ truth|$$

- Cross Entropy Loss (CEL) is the most common function for classification models and calculates loss based on the prediction probabilities. The calculation formula is:

$$CEL = \frac{1}{n} \cdot \sum_{1}^{n} -ln(probability\ of\ ground\ truth\ class)$$

- Multi-class Cross Entropy Loss
- Sparse Multi-class Cross Entropy Loss

The choice of a type of loss function depends very much on the data types and the task solved by the system. Each function has unique properties and allows the algorithm to learn differently for the model to fit the data.

The basic differences between these functions are related to the importance they give to outlier labels among all the labels.

E. Application of parameter adjustment algorithms

The second step in optimizing the model is to adjust the parameters. At this stage, the aim is to minimize the loss in order to obtain the parameters that lead to a model as close as possible to the data. The method used in most of the machine learning algorithms is gradient descent which is a numerical optimization algorithm to find the most suitable weights for the dataset and minimize the loss. The gradient descent algorithm determines the slope and based on it the correct direction for adjusting the parameters so as to reach the minimum value of the loss. The gradient has two characteristics: direction and size. The calculated gradient will indicate the direction and step required to reach the lowest value of the loss function. In the step-by-step approach to the minimum value of the loss by gradient descent, there are a number of factors that are decisive in reaching the minimum and the speed with which it is reached. An example is the learning rate or step size. These factors can be changed manually during the learning process and are called hyperparameters.

F. Running for the validation dataset

The trained model is run on the validation dataset only for loss measurements purposes and hyperparameters adjustments.

G. Calculate loss and save if it the best model yet

If the loss for the validation dataset is the minimum from the so far trained models, then it should be saved before trying new parameters and hyperparameters.

H. Repeat previous steps

The training and validation steps are repeated till there is obtained a well fitted model for the desired application.

I. Choose the best model from the saved ones

The training and validation are finished and the best fitted model is delivered and tested on the test dataset to get the release form performance of the developed application.

6.3 Integrated System for New Product Development

6.3.1 Agile Approach for Healthcare Organizations' Projects

In recent times it's far well-known that now not only companies from the software development industry have located the blessings of adopting Agile approach [1, 2] that suggests extensive collaboration [3], working software over complete documentation, client collaboration and adaptability to adjustments in customer requirements [4].

An adapted Agile approach suits for most healthcare solution projects because it permits the accurate capture of the important capabilities for the end customers (stakeholders-clinical staff or patients) and to supply useful products that address end users' necessities.

In Agile approach the teams work directly with users for design and feedback, IT teams stick with tasks in preference to moving off of projects after preliminary go-live, and budgets and strategic plans are kept as flexible as possible to account for existent uncertainty within the process. SCRUM practices and procedures are being used to effectively control an increasing number of technology projects in healthcare.

SCRUM, as a framework of Agile, gives, besides its conventional utilization in software development, the simplicity of dealing with unpredictability and fixing complex issues, with regard for humans and self-organization and also the precision introduced by time-boxed events, or sprints, managed by the SCRUM team, represented by Product Owner (PO), Scrum Master (SM), and Development Team (DT).

The usage of Agile for projects has enabled several healthcare groups to start to efficiently adapt and adopt innovation in person wellbeing, clinical workflows, and virtual health technology with the ultimate purpose of enhancing care. As an instance, Inception Health and its partnerships had been created to boost up innovation through more speedy cycle project control for guiding innovation, helping its health provider partners in regions of digital transformation, consumerism, and precision medicine [5].

Sindhwani et al. [6] describe the stage of implementation of Agile in healthcare and emphasize that agile approach has to be very well adopted via healthcare organizations to satisfy the customers' needs. Boustani et al. [7] describe the critical components of an Agile implementation, which fast and effectively implements evidence-based healthcare solutions, and present a case by demonstrating its application.

The shift to agile healthcare businesses has also been addressed by Patri and Suresh [8] showing the enablers of Agile performance in healthcare organizations. Kitzmiller et al. [9] underlined the necessity to adapt traditional plan-driven implementation strategies to contain agile techniques. Also, the Agile approach proved to be a strong exponent of change in eHealth innovation systems [10]. Agile Innovation was brought as a method any complex adaptive organization can adopt to acquire fast, systematic, client-centered development of innovative interventions [11].

Agile achievement stories of leading healthcare businesses as GE and Siemens that built products in the clinical imaging area, Philips that developed a healthcare management systems and clinical devices and as Abbott that advanced a blended hardware/software medical device have been reported by Birk [12].

Additionally, the potential advantages of integrating particular Agile methodologies in healthcare IT projects have been addressed by Goodison et al. [13] and the findings revealed there's scarce literature available on agile project management methodologies used in healthcare IT systems implementations.

McCaffery et al. [14] examined the challenges and spotlighted agile practices that have been adopted with success within the medical device software industry. Rottier and Rodrigues [15] provide the case of a clinical device corporation that adapted SCRUM and mapped it into the process to be able to satisfy the strict standards and regulations. Košinár [16] deliver a technique to modeling of software strategies for healthcare management systems development on SCRUM Agile approach and formal modeling tools.

The advent of the SCRUM methodology from Agile approach component in the integrated system is supported by the current trend of organizations in the healthcare and medical device software industry.

6.3.2 Model Based on the Modified QFD Method for Agile SDLC

Considering the QFD state-of-the-art [17, 18], there's a strongly acceptance that there are no fields wherein the QFD method was not implemented, tried to be implemented or impossible to be implemented and basically, there is no definite boundary for QFD's possible fields of applications [19]. So, based on a scientific literature evaluation, Sukma et al. [20] provide a wide understanding and expertise of the application of QFD within the healthcare industry, showing that QFD can be used to enhance service and create customer satisfaction. QFD applications also are frequently incorporated with other approaches in order that QFD can be stated as a complementary method that is able to provide maximum results.

Schockert and Herzwurm [21] introduce Agile software QFD characterized with the aid of the embedding in the agile iteration cycle and specific methodological features along with the incrementally growing prioritization matrix and the priority map. Both Agile software development and Software QFD gain from Agile Software QFD, being the expression of a genuinely business value oriented, agile requirements engineering embedding QFD in an iterative and incremental development technique.

The model primarily based at the modified QFD approach for Agile SDLC is the result of the QFD integrated model for projects using the Agile method [22].

This model of the integrated system is based on the modified QFD to the SCRUM methodology so as to make a contribution to embracing innovation in the patient wellbeing and also, the quantification of the impact of modifications within the NPD at any moment with the help of the computed index (Offset), representing the proportion of achievement for the product on the current stage. These changes could have widespread implication, specifically when several constraints exist (like, time, fitting into a budget or the availability of the human resource), through introducing some risks that need to be acknowledged and properly controlled on the way to not jeopardizing the new product development.

For the case of NPD using the SCRUM methodology of the Agile approach, the house of quality was adapted, as presented in Figs. 6.4 and 6.5.

WHAT section refers to gathering the user stories based on client demands. User stories are basically the translation of clients demands in a way that can help determine WHAT the client wants and WHY it wants it, or other put, what value does that requirement bring to the client.

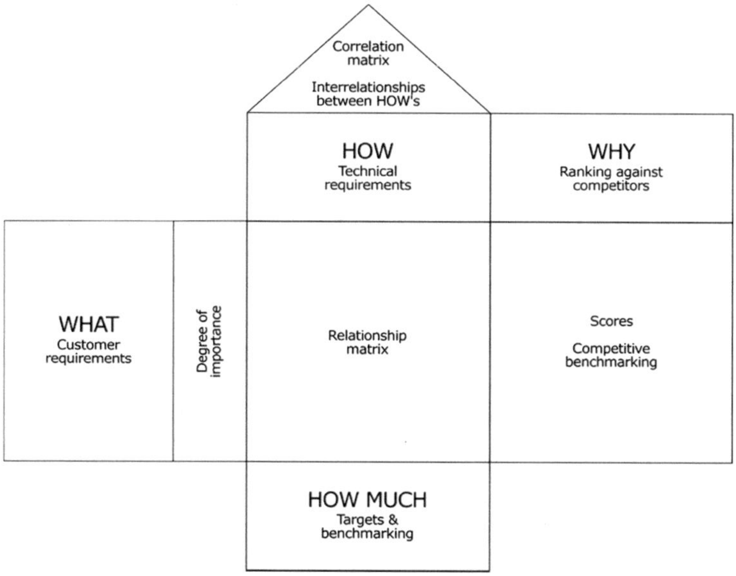

Fig. 6.4 House of quality. Adapted [23]

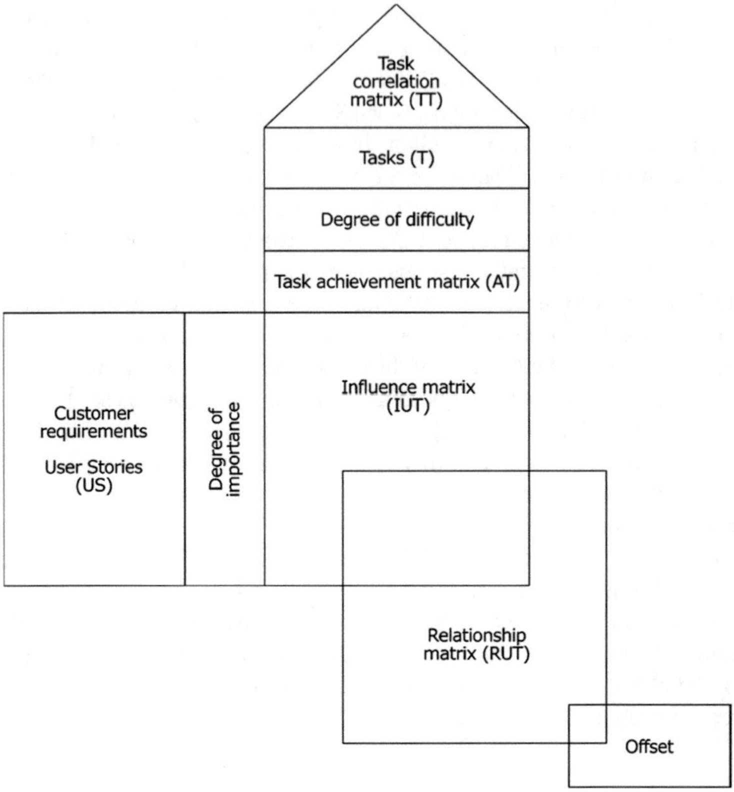

Fig. 6.5 House of quality adapted for SCRUM projects

HOW section refers to filling in to the Tasks segment that contains the tasks to be completed in order to develop the product desired by the stakeholder.

The tasks are generally deducted by the Scrum Master from the User Stories segment. But, in the case of healthcare products, the link between user stories and task, i.e. the WHAT and the HOW, is difficult to achieve, due to the complexity and the necessity to analyze big amounts of data. This is a good job for machine learning, the WHY.

HOW MUCH—the offset index value that shows the level (percentage) of accomplishment for the product at the current moment.

The calculation of the index started from the specific mechanism of the QFD method of using the matrices, but these were adapted to the SCRUM methodology.

The Offset index is calculated using the following formula:

$$offset = \sum_{j=1}^{m}\sum_{i=1}^{n}\frac{RUT(i,j)\cdot US(i)}{\sum_{i=1}^{n}US(i)}$$

where

m—represents the number of work tasks.

n—represents the number of customer requirements.

RUT—represents the values in the Relationship Matrix, $RUT(i, j)_{i=\overline{1,n}, j=\overline{1,m}}$, the matrix in which the actual functional values obtained by performing certain pre-selected work tasks are recorded, $RUT(i, j) \in [0, 100\%]$.

US—the degree of importance of customer requirements, $US(i)_{i=\overline{1,n}}$.

The values in the Relationship Matrix are calculated using the following formula:

$$RUT(i, j) = \begin{cases} 0, if \prod_{k=1}^{m}(AT(k) + 1 - TT(k, j)) = 0 \\ \frac{IUT(i,j) \cdot T(j) \cdot AT(j)}{\sum_{r=1}^{m} IUT(1,r) \cdot T(r)}, else \end{cases}$$

where

AT—represents the Task Achievement stage, $AT(j)_{j=\overline{1,m}}$, which can have either value 0 (for an unfinished task) or 1 (for a completed task).

TT—represents the Task Correlation Matrix, $TT(i, j)_{i,j=\overline{1,m}}$, in which the interdependence relations between the work tasks are highlighted, this can take the value 0 if there is no inhibition correlation between tasks or 1 if task i inhibits task j.

IUT—represents the Influence Matrix, $IUT(i, j)_{i=\overline{1,n}, j=\overline{1,m}}$, a matrix that reflects the share of each task in a customer requirement, with $IUT(i, j) \in [0, 100\%]$, and $\sum_{j=1}^{m} IUT(i, j) = 100\%; with\, i = \overline{1, n}$

T—the degree of difficulty of each task, $T(j)_{j=\overline{1,m}}$, usually expressed as time to achieve.

The computation the Offset index requires the following steps: Identifying and encoding customer requirements in a form that incorporates the core functionality pursued by the customer; Identifying the importance of each customer requirement, in order to identify requirements that need to be met immediately and those that can be met later. Based on the identified requirements, there are then deduced, using machine learning, the tasks needed to be completed, their degree of difficulty and the interdependencies between them. Then it is set out the extent to which the requirements are covered by the established tasks, in a percentage form, within the IUT matrix. The fulfillment of the tasks is verified during product development, recording the completed tasks with 1 and the uncompleted ones with 0 in the AT matrix. Based on the recorded input information, the values within the RUT matrix can be calculated and then the Offset index can also be determined.

6.3.3 Optimal Tasks Planning Algorithm

Nowadays there are different online project management platforms that are currently being used to plan tasks and monitor their performance. Although these platforms have multiple functionalities and provide a battery of tools useful for project management, none of these platforms provide an automatic way to divide tasks, but only allow the introduction of these tasks and set certain parameters for them, such as: assignment of a task to a specific person, setting a deadline for the completion of the task, determining the degree of importance or difficulty, etc.

Regarding the current ways of dividing and prioritizing work tasks, there are several approaches used in practice and recognized in the literature, such as: Cumulative Voting, MoSCoW Analysis, Eisenhower Matrix or focusing on the decision-making power of the highest paid person from the company (HiPPO—Highest-paid person's opinion).

Another variant of prioritizing work tasks is based on identifying the value to be brought, as follows:

- Prioritizing tasks according to the value brought to the client;
- Prioritizing tasks according to the value brought to the business;
- Choosing tasks that can be performed immediately to the detriment of the most complex;
- Choosing the riskiest tasks to start with;
- Choosing tasks by calculating the high costs of not fulfilling them;
- Choosing tasks by identifying the interdependencies between them and following the logical thread of product development;
- Choosing the tasks that contribute the most to achieving the goal set for the current development stage.

In the context of methods of prioritization and efficient division of tasks, the proposed algorithm combines three variants of prioritization of tasks, namely: the choice of tasks according to the interdependencies between them, the choice of those that contribute most to achieving the goal established for the current stage of development and prioritization of tasks according to the value brought to the client.

The algorithm aims to obtain an optimal and objective result in terms of selecting tasks and dividing them into development periods called development stages. In addition, the algorithm reduces the time spent planning the tasks included in each required step, and moreover, provides an objective solution, devoid of the subjectivity of the members of the development team and their interests.

Another problem solved by the algorithm is the provision of a tool that can quantify the degree of fulfillment of customer requirements at a given time, taking into account their rapid change. In this sense, the targeted solution aims to integrate an iterative approach to product development requested by customers. The solution is based on a production process approach based on the PDCA (Plan, Do, Check, Act) continuous improvement cycle to help organize and carry out management activities.

Compared to traditional product development methods in which customer requirements are identified, the required products are developed and then delivered to the customer at the end of the development period, in the proposed approach, after identifying customer requirements, the product development team identifies the most important aspects of the desired products and develops them, delivering to the customer real functionalities of the requested product, at predetermined time intervals. This helps the developer to tailor their products to customer requirements, even if they require significant changes to the shape or functionality of the desired product. This method allows the development team to ensure that the product developed is the one desired by the customer at every stage of the development process.

6.3.4 NPD Based on Machine Learning

The integration of machine learning into new product development has emerged as a necessity given by the demand for healthcare products and services of greater complexity due to the increasing and diverse requirements of stakeholders that must be properly managed to meet them.

The system is considered a new product development from a functional point of view. The system is built starting from a predetermined number of inputs u_1, u_2,..., u_n and outputs y_1, y_2,..., y_m.

The inputs u_i can be measured directly, indirectly or attributively and are obtained from potential users in a pilot group sized according to the data requirements for the appropriate machine learning algorithm. They must provide useful and sufficient data based on experience. The most appropriate way to obtain this data is identified, strictly considering the functionality of the new product.

The outputs y_j are actions produced by the system, such as prediction, classification, or control signal to a device.

The system is a black box with the functionality characterized by User stories. For the use of machine learning in black box it is important to choose the appropriate type of model, depending on the specifics of the system determined by user stories. The role of machine learning is to learn the desired behavior of the system (Fig. 6.6).

From the above, results the functional tasks for the development of the system. To these are added design tasks extracted directly from user stories.

Although it is the patient who benefits from healthcare systems, the client of healthcare applications is not the patient. The client of the applications is the specialist for whom the application offers support and augmentation in his clinical activity. This means that no successful healthcare application can be developed without the substantial involvement of medical professionals. So, they will be the main actors in formulating the requirements. The requirements can be deduced in several ways, including: direct discussions with the customer about the desired product or service, market research, opinion polls, reports from the medical stakeholders, from existing standardized medical devices, etc.

Based on the requirements, the system/application is designed using machine learning.

The most important things to keep in mind are the available data and the right type of machine learning algorithm. Regarding the data, we have two variants: labeled data, in which case we will perform supervised machine learning and untagged data,

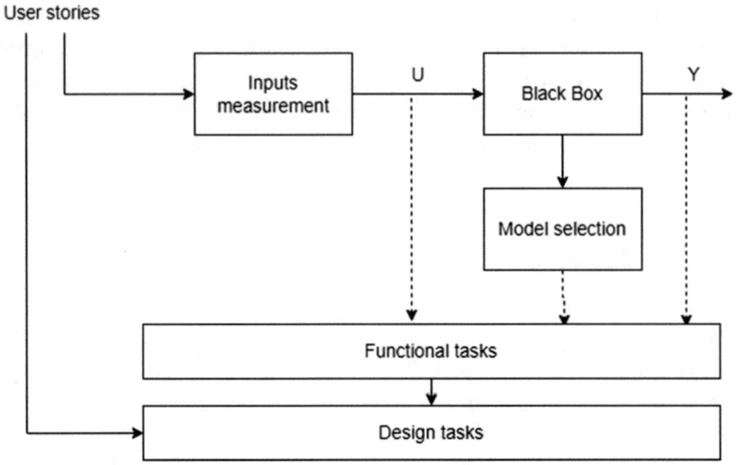

Fig. 6.6 Tasks extraction

in which case we will perform unsupervised machine learning. The algorithms are chosen based on the type and size of existing data.

In this way we have the core of the system, thus being solved the functionality and resulting the tasks for the implementation in the final machine learning system developed. To these tasks are added those related to product design that take into account both the requirements expressed by the user stories and the functional principles related to the machine learning algorithm. In the case of the development of medical devices (hardware + software), the design tasks also include the choice of sensors compatible with those in the medical devices with which the data were collected, as well as the related actuating elements, if applicable.

After determining the tasks, there is available an overview of the entire product, which allows the establishment of the correlations between tasks and the difficulty of each task expressed in time units. In addition, both functional and design tasks are determined from the requirements expressed in the user stories, a process that results in the relationships between tasks and requirements, thus providing the inputs for the algorithm (Fig. 6.7).

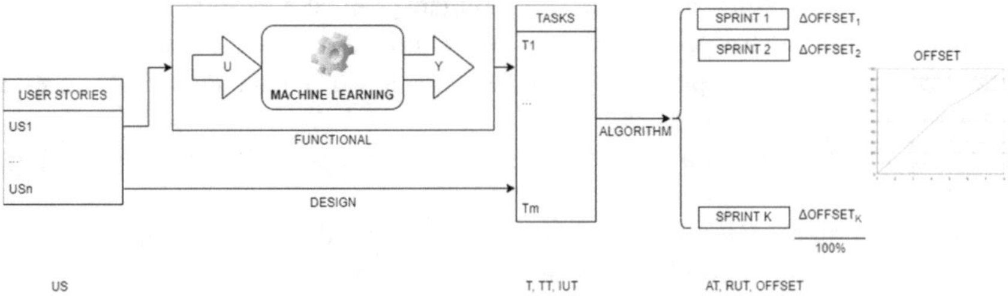

Fig. 6.7 Integrated system block diagram

The algorithm automatically prioritizes tasks in development stages and uses all input data to determine the tasks needed to be completed for each stage of the development process, its logic diagram being shown in Fig. 6.8.

The algorithm is based on the idea of dividing the tasks into development stages in a way that allows both maximizing customer satisfaction and the functionality obtained at the end of each development stage, the satisfaction being measured using the Offset index.

The logic diagram shown in Fig. 6.8 presents how tasks are selected for distribution in product development stages. Thus, initially the objective duration of product development is declared, in number of sprints, which results in the variable sprint_offset [calculated as 100/no. sprints], a variable that represents the value of the Offset index for the current development stage, and the TS vector, which contains the tasks selected in the current stage.

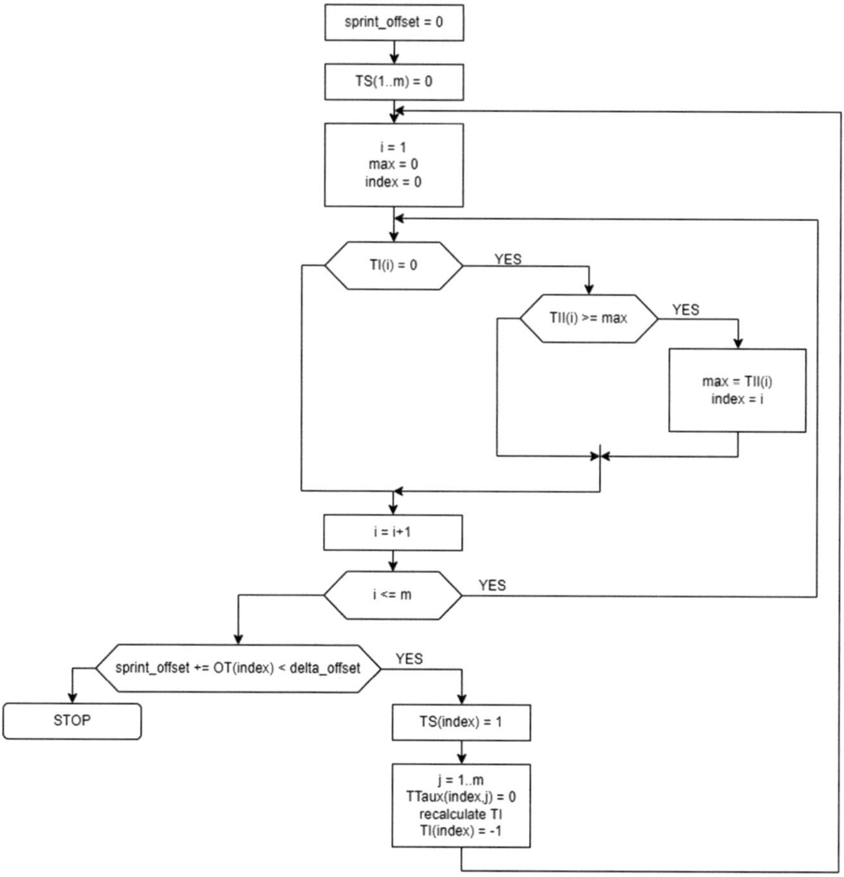

Fig. 6.8 Logic diagram

Next there can be see the repetitive loop characteristic of an algorithm for determining the maximum in a series of values, initiated by declaring the iterator i and the variables max and index that have the role of counting the maximum number of tasks inhibited by an uninhibited task, respectively its index; the initial conditional block ensures the execution of the maximum algorithm taking into account only the uninhibited tasks; the next conditional block acts as a comparator between the number of tasks inhibited by the current iteration-specific task and the value of the max variable, directing execution so that when a value greater than or equal to that of the max variable is triggered, the assignment block is triggered whereby the value of the max variable is overridden by the number of tasks inhibited by the current task and the index variable records the current value of iterator i, ie the actual index of the corresponding task; by incrementing the iterator i and querying the condition of stopping the repetitive loop, its execution is ensured in the interval [1, m], where m represents the number of existing tasks.

After determining the uninhibited task that inhibits most other tasks, the value of the Offset index obtained by performing that task is recalculated. By comparing this value with that of the DeltaOffset index, the stopping condition of the outer repetitive loop underlying the recursive algorithm for distributing tasks to development stages is established. The DeltaOffset index shows the actual value of the Offset index in the current development period, by calculating the difference between the value of the Offset indicator at the end and its value at the beginning of the development period.

As long as the cumulative value of the maximum number of tasks inhibited by the previously identified tasks does not exceed the value of the DeltaOffset index, the algorithm records the index of the task identified in the TS vector, then invalidates the task row in the copy auxiliary of the TTaux Correlation Matrix by overwriting each element with the value 0, thus considering that the selected task has been completed and no longer inhibits other tasks.

As a preliminary step to the next iteration, the TI vector, which contains the tasks left unallocated to the iterations, is recalculated to determine the new list of uninhibited tasks (from which, again, the one that inhibits most other tasks will be selected). It is also necessary to invalidate the current index position of the TI vector in order to avoid its evaluation in the next executions of the algorithm for determining the maximum inhibited tasks. The algorithm will run until the cumulative value of the sprint_offset variable exceeds the value of the DeltaOffset index, when the execution is finished and it is possible to inspect the tasks assigned to the current work step by analyzing the positions of the TS vector with value 1.

Advantages of using the integrated system for NPD

- Choosing tasks/sprints in such a way that the developed functionality is also visible
- Determining at each step the amount of functionality added by DeltaOffset and the total functionality by Offset index
- Real-time updating of the tasks related to each sprint based on the data regarding their accomplishment
- Possibility to simulate product development in sprints

- Can be applied to any product either pure software (such as imaging) or hardware and software (such as medical device).

The particularization for healthcare was made through the stage of analyzing the functionality of the system as a machine learning type system. The result of product development using this integrated system is a functional prototype that can be used for mass production after the completion of mandatory medical certifications.

6.4 NPD Results in the Integrated System Context

We present two studies of the design of new healthcare products with different approaches related to their functionality.

Thus, the first example is an exoskeleton device designed for people with upper limb mobility problems [24, 25]. For this case, the machine learning algorithm is trained to transmit control signals for the movement of the exoskeleton torques based on the information gathered from the sensors.

The second example is an early burnout detection device [26–29] in which the machine learning algorithm estimates the person's condition based on information gathered from sensors and signals the detected state.

6.4.1 Exoskeleton

The model is implemented starting with understanding and describing the customer requirements or needs both from a functional and a design point of view, taking into account that this is a device that is worn by the patient.

Since the development model is based on the SCRUM framework, the requirements take the form of five user stories (US), each with its own degree of importance, on a scale from 1 to 10. Based on this, the PO together with the SM will decided the order and the timeframe for the customer requirements. The tasks for achieving those requirements are determined by applying a selected machine learning algorithm, linear regression, with data from EMG (Electro Myography) and IMU (Inertial Measurement Unit) sensors as inputs and arm position as output, using supervised learning, and by analyzing the user stories from design point of view (Fig. 6.9).

For this case, resulted 26 tasks, each with a level of difficulty expressed in achievement time, by the SM together with the DT.

Then, based on the user stories and the resulted tasks, is determined how much of a US is covered by each task. Also, there are established the correlations between the resulted tasks.

Having all these information, the tasks prioritization algorithm will provide the sprints content in best fitted tasks in order to develop the new product and have at each step maximum visible functionality.

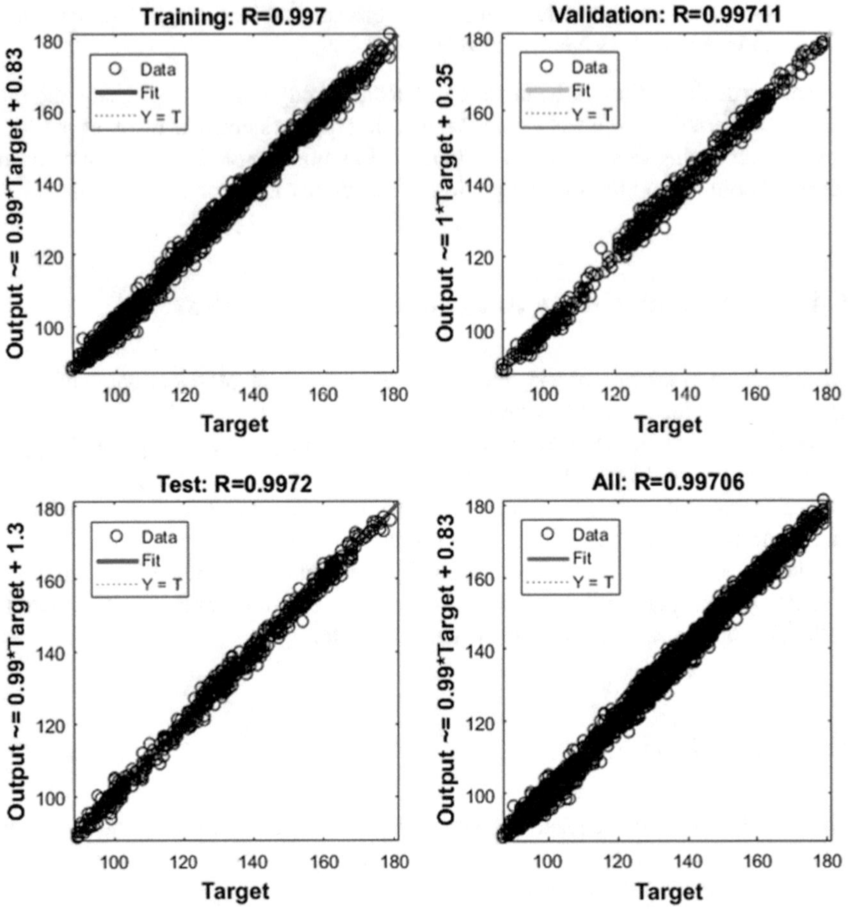

Fig. 6.9 Linear regression training

There was decided a period of 6 weeks for the development of the project and also the decision was to divide the work in 6 sprints of 7 days each. This decision to divide the tasks in short iterative development periods with an output at the end of each iteration is in accordance to the Agile principles that encourage rapid and iterative releases.

Considering an ideal linear continuous development rate, the expected growth per sprint was 16.67%, expressed by the offset index values. The real growth rate is presented in Fig. 6.10, and the real growth rate is higher in the first half of the period of the project and decreases in the second half.

The algorithm for automatic task prioritization has as main objective the optimization of the decision-making process of selecting the task or tasks that would be best to be completed for each sprint.

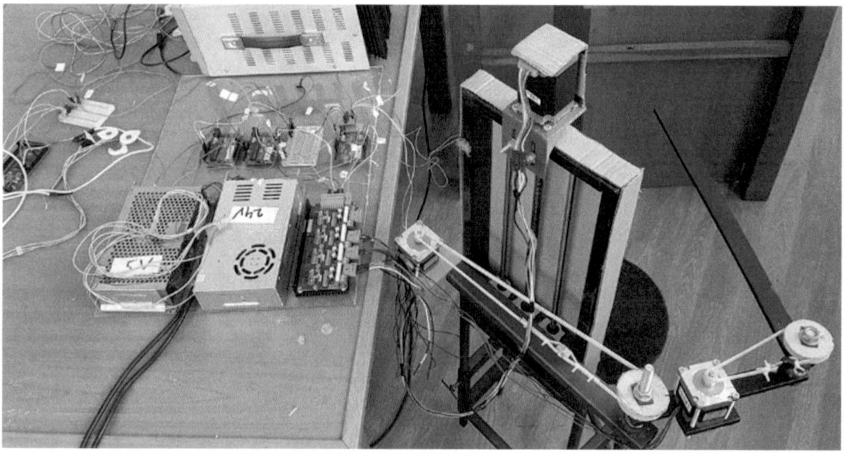

Fig. 6.10 The developed exoskeleton prototype

The results of applying the algorithm with the help of Octave environment were as follows: First sprint: tasks—2, 6, 7, 8, 9 and 15; Second sprint: tasks—3, 10, 11, 18, 19 and 20; Third sprint: tasks 1, 4, 5, 22, 23, 24 and 25; Fourth sprint: tasks 12, 13 and 14; Fifth sprint: tasks 16, 17 and 21; Sixth sprint: task 26.

The average growth rate for each sprint was 23% for the first half of sprints and 10% for the second half, cumulating to a value of 100% at the end of the last sprint.

The version 1 of the exoskeleton prototype resulted from applying the integrated system for new product development is presented in Fig. 6.10 and the growth of the offset indicator value is also graphically represented in Fig. 6.11.

6.4.2 Burnout Device

The model is implemented starting with understanding and describing the customer requirements or needs. Since the model is based on the SCRUM framework, the requirements take the form of eight user stories (US), each with its own degree of importance, on a scale from 1 to 10. Based on this, the PO together with the SM will decided the order and the timeframe for the customer requirements. The tasks for achieving those requirements are determined by applying a selected machine learning algorithm, multi-class classification, with data from Heart Rate and Oximetry sensors together with information regarding the organizational climate as inputs and stress level as output, using supervised learning, and by analyzing the user stories from design point of view (Fig. 6.12).

For this case, resulted 15 tasks, each with a level of difficulty expressed in achievement time, by the SM together with the DT.

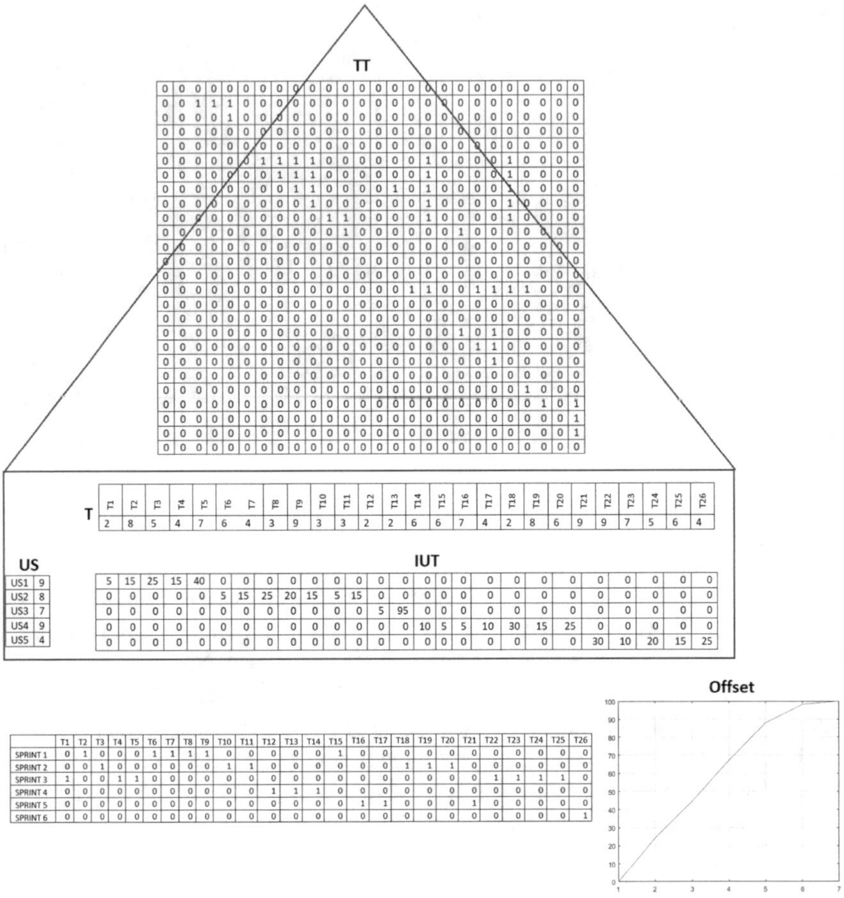

Fig. 6.11 Prioritization algorithm for exoskeleton device development

Then, based on the user stories and the resulted tasks, is determined how much of a US is covered by each task. Also, there are established the correlations between the resulted tasks.

Having all these information, the tasks prioritization algorithm will provide the sprints content in best fitted tasks in order to develop the new product and have at each step maximum visible functionality.

There was decided a period of 8 weeks for the development of the project and also the decision was to divide the work in 7 sprints of 8 days each. This decision to divide the tasks in short iterative development periods with an output at the end of each iteration is in accordance to the Agile principles that encourage rapid and iterative releases.

Considering an ideal linear continuous development rate, the expected growth per sprint was 12.5%, expressed by the offset index values. The real growth rate

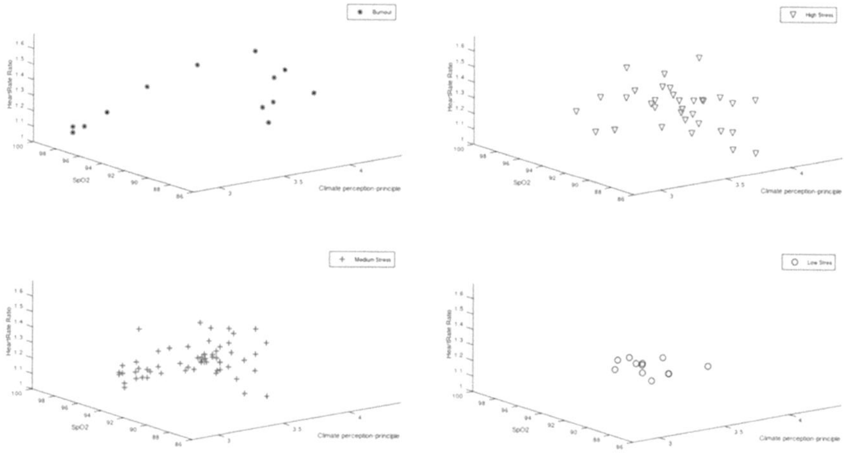

Fig. 6.12 Logistic regression training

is presented in Fig. 6.13, and the real growth rate is higher at the beginning of the project and decreases at the end.

The algorithm for automatic task prioritization has as main objective the optimization of the decision-making process of selecting the task or tasks that would be best to be completed for each sprint.

The results of applying the algorithm with the help of Octave environment were as follows: First sprint: tasks—1, 2, 3, 4, 5 and 9; Second sprint: tasks—6 and 10; Third sprint: task 14; Fourth sprint: task 8; Fifth sprint: tasks 7 and 12; Sixth sprint: task 15; Seventh sprint: tasks 11 and 13.

The average growth rate for each sprint was 14.28% cumulating to a value of 100% at the end of the last sprint. The growth of the offset indicator value is also graphically represented in Fig. 6.13.

The version 1 of the burnout estimation device prototype resulted from applying the integrated system for new product development is presented in Fig. 6.14.

6.5 Conclusions

The chapter provides the application of an integrated system for new healthcare product development, that has a theoretical foundation based on machine learning, Agile and modified QFD method. All these elements are presented with their mathematical support. The chapter involves practical implications proved by the experimental results for developing two innovative healthcare products, one for an exoskeleton for upper limb, useful to persons with low arm mobility, and the other

TT

0	0	1	1	0	0	1	1	0	0	0	0	0	1	0
1	0	1	1	1	0	1	0	1	1	1	0	1	1	0
0	0	0	1	1	1	1	0	0	0	0	0	0	1	0
0	0	0	0	0	0	1	0	0	0	0	0	1	1	0
0	0	0	0	0	1	0	0	0	1	1	0	0	1	0
0	0	0	0	0	0	0	0	0	0	0	1	0	1	0
0	0	0	0	0	0	0	0	0	0	0	0	0	0	0
0	0	0	0	0	0	0	0	0	0	0	0	0	0	1
0	0	0	0	0	0	0	0	0	1	1	0	1	0	0
0	0	0	0	0	0	0	0	0	0	1	0	1	0	0
0	0	0	0	0	0	0	0	0	0	0	0	1	0	0
0	0	0	0	0	0	0	0	0	0	0	0	0	0	0
0	0	0	0	0	0	0	0	0	0	1	0	0	0	0
0	0	0	0	0	0	0	1	0	0	0	0	0	0	1
0	0	0	0	0	0	0	0	0	0	0	0	0	0	0

T

T1	T2	T3	T4	T5	T6	T7	T8	T9	T10	T11	T12	T13	T14	T15
2	3	8	6	4	2	10	5	7	7	9	1	5	6	6

US

US1	9
US2	8
US3	8
US4	7
US5	7
US6	4
US7	8
US8	5

IUT

20	20	20	10	5	5	15	5	0	0	0	0	0	0	0
0	0	0	0	0	0	0	0	40	25	35	0	0	0	0
0	0	0	0	0	50	0	0	0	0	0	50	0	0	0
0	0	0	0	0	0	0	0	0	25	25	0	50	0	0
0	0	0	0	0	0	0	30	0	0	0	0	0	50	40
0	0	0	0	0	0	0	0	0	0	0	0	0	0	100
0	0	0	0	0	0	0	60	0	0	0	0	0	40	0
0	0	0	0	0	0	0	60	0	0	0	0	0	40	0

	T1	T2	T3	T4	T5	T6	T7	T8	T9	T10	T11	T12	T13	T14	T15
SPRINT 1	1	1	1	1	1	0	0	0	1	0	0	0	0	0	0
SPRINT 2	1	1	1	1	1	1	0	0	1	1	0	0	0	0	0
SPRINT 3	1	1	1	1	1	1	0	0	1	1	0	0	0	1	0
SPRINT 4	1	1	1	1	1	1	0	1	1	1	0	0	0	1	0
SPRINT 5	1	1	1	1	1	1	1	1	1	1	0	1	0	1	0
SPRINT 6	1	1	1	1	1	1	1	1	1	1	0	1	0	1	1
SPRINT 7	1	1	1	1	1	1	1	1	1	1	1	1	1	1	1

$$RUT(i,j) = \begin{cases} 0, & if \prod_{k=1}^{m}\left(AT(k) + 1 - TT(k,j)\right) = 0 \\ \frac{IUT(i,j) * T(j) * AT(j)}{\sum_{r=1}^{m} IUT(1,r) * T(r)}, & else \end{cases}$$

$$offset = \sum_{j=1}^{m}\sum_{i=1}^{n}\frac{RUT(i,j) * US(i)}{\sum_{i=1}^{n} US(i)}$$

SPRINT 1	SPRINT 2	SPRINT 3	SPRINT 4	SPRINT 5	SPRINT 6	SPRINT 7
15.60	32.05	47.80	64.18	73.53	85.02	100.00

Offset

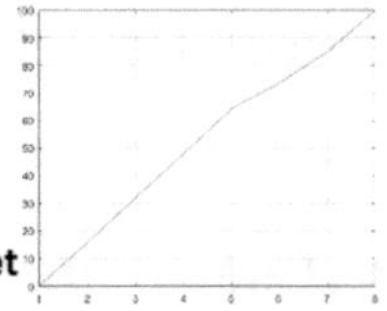

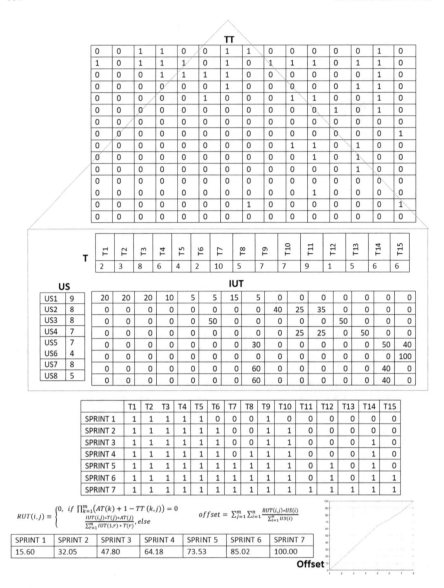

Fig. 6.13 Prioritization algorithm for burnout estimation device development

for a burnout early detection device, useful for alarming the occurrence of chronical stress state, related to the organizational climate and based on the physiological parameters.

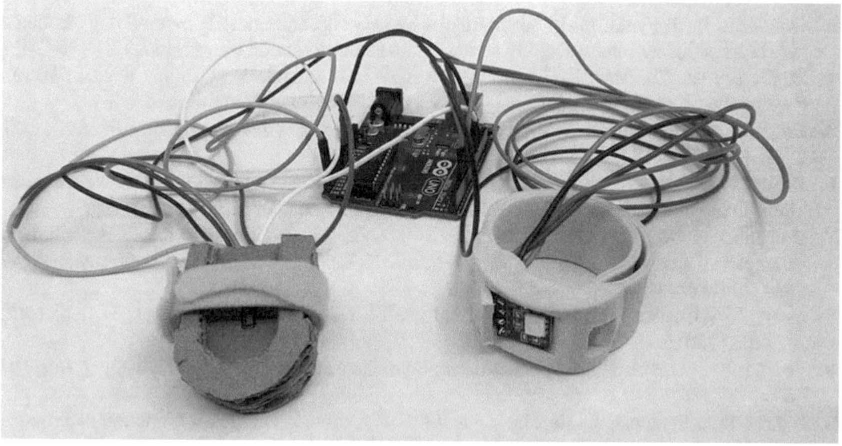

Fig. 6.14 The developed burnout prototype

References

1. Campanelli, A.S., Parreiras, F.S.: Agile methods tailoring—a systematic literature review. J. Syst. Softw. **110**, 85–100 (2015)
2. Salah, D., Paige, R.F., Cairns, P.: A systematic literature review for Agile development processes and user centered design integration. In: Proceedings of the 18th International Conference on Evaluation and Assessment in Software Engineering EASE, vol. 14, pp. 1–10 (2014)
3. Inayat, I., Salim, S.S., Marczak, S., Daneva, M., Shamshirband, S.: A systematic literature review on Agile requirements engineering practices and challenges. Comput. Hum. Behav. **51**(B), 915–929 (2014)
4. Beck, K., Beedle, M., van Bennekum, A., Cockburn, A., Cunningham, W., Fowler, M.: Manifesto for Agile Software Development [Online] (2001). http://agilemanifesto.org/
5. Crotty, B.H., Somai, M., Narath Carlile, N.: Adopting Agile Principles in Health Care [Online] (2019). https://www.healthaffairs.org/do/10.1377/forefront.20190813.559504/full/
6. Sindhwani, R., Singh, P.L., Prajapati, D.K., Iqbal, A., Phanden, R.K., Malhotra, V.: Agile system in health care: literature review. In: Shanker, K., Shankar, R., Sindhwani, R. (eds.) Advances in Industrial and Production Engineering. Lecture Notes in Mechanical Engineering, pp. 643–652 (2019). https://doi.org/10.1007/978-981-13-6412-9_61
7. Boustani, M., Alder, C.A., Solid, C.A.: Agile implementation: a blueprint for implementing evidence-based healthcare solutions. J. Am. Geriatr. Soc. **66**(7), 1372–1376 (2018). https://doi.org/10.1111/jgs.15283
8. Patri, R., Suresh, M.: Modelling the enablers of Agile performance in healthcare organization: a TISM approach. Glob. J. Flex. Syst. Manag. **18**(3), 251–272 (2017)
9. Kitzmiller, R., Hunt, E., Breckenridge-Sproat, S.: Adopting best practices CIN: computers, informatics. Nursing **24**(2), 75–82 (2006)
10. Velthuijsen, H., Balje, J., Carter, A.: Agile development as a change management approach in healthcare innovation projects. Paper presented at 3rd Understanding Small Enterprises (USE) Conference 2015, Groningen, Netherlands, p. 16 (2015)
11. Holden, R.J., Boustani, M.A., Azar, J.: Agile innovation to transform healthcare: innovating in complex adaptive systems is an everyday process, not a light bulb event. BMJ Innov. **7**, 499–505 (2021)
12. Birk, A.: Agile Success Stories in Healthcare. SWPM White Paper. ID 2021-03, V1.1:7 (2021)

13. Goodison, R., Borycki, E.M., Kushniruk, A.W.: Use of Agile project methodology in health care IT implementations: a scoping review. Stud. Health Technol. Inform. **257**, 140–145 (2019)
14. McCaffery, F., Trektere, K., Top, Ö.Ö.: Agile—is it suitable for medical device software development? In: 16th International Conference, SPICE 2016 Proceedings (2016)
15. Rottier, P.A., Rodrigues, V.: Agile development in a medical device company. In: Agile 2008 Conference, pp. 218–223 (2008)
16. Košinár, M.: Knowledge modeling of Agile processes in healthcare systems development biomedical engineering and informatics. Clin. Techonol. **43**(4), 32–35 (2013)
17. Rothenberger, M., Kao, Y.C., Wassenhove, L.N.V.: Total quality in software development: an empirical study of quality drivers and benefit in Indian software projects. Inf. Manag. **47**, 372–379 (2010)
18. Yong, J., Wilkinson, A.: Rethinking total quality management. Total Qual. Manag. **12**(2), 247–258 (2001)
19. Chan, L.K., Wu, M.L.: Quality function deployment: a literature review. Eur. J. Oper. Res. **143**, 463–497 (2002)
20. Sukma, D.I., Setiawan, I., Kurnia, H., Atikno, W., Purba, H.H.: Quality function deployment in healthcare: systematic literature review. J Sist Tek. Ind. **24**(1), 15–27 (2022)
21. Schockert, S., Herzwurm, G.: Agile software quality function deployment. In: 23rd International QFD Symposium, ISQFD 2017 JUSE (2017)
22. Ionica, A., Leba, M., Dovleac, R.: A QFD based model integration in Agile software development. In: 12th Iberian Conference on Information Systems and Technologies (CISTI), pp. 1–6 (2017). https://doi.org/10.23919/CISTI.2017.7975995
23. Govers, C.P.M.: What and how about quality function deployment (QFD). Int. J. Product. Econ. (46–47), 575–585 (1996)
24. Risteiu, M.N., Leba, M.: Right Arm Exoskeleton for Mobility Impaired. Advances in Intelligent Systems and Computing, vol. 1160, pp. 744–754. Springer (2020). ISBN 978-3-030-45687-0. ISBN 978-3-030-45688-7
25. Risteiu, M., Leba, M., Stoicuta, O., Ionica, A.: Study on ANN based upper limb exoskeleton. In: IEEE 20th Mediterranean Electrotechnical Conference (MELECON), pp. 402–405 (2020). https://doi.org/10.1109/MELECON48756.2020.9140691
26. Dovleac, R., Ionica, A., Leba, M.: QFD embedded Agile approach on IT startups project management. Cogent Bus. Manag. **7**(1) (2020)
27. Dovleac, R., Risteiu, M., Ionica, A.C., Leba, M.: Mobile burnout estimation device—an Agile driven pathway. In: Rocha, Á., Adeli, H., Dzemyda, G., Moreira, F., Ramalho Correia, A.M. (eds.) Trends and Applications in Information Systems and Technologies. WorldCIST 2021. Advances in Intelligent Systems and Computing, vol. 1365, pp. 522–531 (2021)
28. Leba, M., Ionica, A.C., Nassar, Y., Riurean, S.: Machine learning approach on burnout—the case of principals from Southern Israel. In: Antipova, T. (ed.) Comprehensible Science. ICCS 2021. Lecture Notes in Networks and Systems, vol. 315, pp. 14–20 (2022)
29. Riurean, S., Leba, M., Ionica, A., Nassar, Y.: Technical solution for burnout, the modern age health issue. In: IEEE 20th Mediterranean Electrotechnical Conference (MELECON), pp. 350–353 (2020). https://doi.org/10.1109/MELECON48756.2020.9140516

Chapter 7
Towards Energy Efficient Data Centers and Computation: Exploring Some Ideas from Physicist's Perspective

Florentin Smarandache, Victor Christianto, and Yunita Umniyati

Abstract As it is known that energy efficient computation has become one of primary criteria for computer system design, especially when it comes to designing large-scale data centers or networks. While such data center best practices have been published, it is still considered normal that around 40% of energy usage at such data centers arise as result of cooling system. Therefore one way to reduce energy usage is to consider energy efficient cooling system. In this paper we consider energy efficient data centers both in four season countries and in two seasons countries. Apart of those data centers, another open problem is how to reduce energy use in P2P network, and we discuss these issues shortly too. These ideas merely reflect physicist's perspective on this subject.

Keywords Energy efficient computation · Computer system design · Data centers · Cooling system · P2P network · Cool computation

7.1 Introduction

As information systems including storage and retrieval systems have grown almost exponentially in recent decades, there is growing concern on energy usage of those information technology and network systems. And most of those IT systems use client–server architecture, which means that most energy burdens come to large data centers, which have grown to become large server farms.

F. Smarandache
Mathematics, Physical and Natural Sciences Division, University of New Mexico, Gallup, NM 87301, USA
e-mail: smarand@unm.edu

V. Christianto (✉)
Malang Institute of Agriculture, Malang, Indonesia
e-mail: victorchristianto@gmail.com

Y. Umniyati
Department of Mechatronics, Swiss-German University, Tangerang, Indonesia
e-mail: yunita.umniyati@sgu.ac.id

L. Ivascu et al. (eds.), *Intelligent Techniques for Efficient Use of Valuable Resources*, Intelligent Systems Reference Library 227,
https://doi.org/10.1007/978-3-031-09928-1_7

Data centers are becoming a significant energy consumer. Server workload, cooling, and supporting infrastructure represents large loads for the grid. Estimates by 2010 tells us that data centers consume up to 2% of total electricity use in USA, and around 3% of total electricity use worldwide [2].

While such data center best practices have been published, it is still considered normal that around 40% of energy usage at such data centers arise as result of cooling system. Therefore, one way to reduce energy usage is to consider energy efficient cooling system. In this paper we consider energy efficient data centers both in four season countries and in two seasons countries. Apart of those data centers, another open problem is how to reduce energy use in P2P network, and we discuss these issues shortly too. These ideas merely reflect physicist's perspective on this subject.

Although this is not an extensive literature review, allow us to discuss some ideas. Some of these ideas are technical but some are not so heavily in technical speaking.

First of all, let us discuss our approach in considering ideas which may work or not in practical system design.

7.2 The Role of Neutrosophic Logic in Exploring New Ideas

It has become part of our belief, that any effort to depict or map life or reality as an abstract substance needs to use real life or concrete experience to arrive at such an understanding. To choose the practical experience and to connect it with the abstract domain, one needs intuition.

As we emphasize on our paper [3]:

> More "right brain" activity, based on direct experiences, leads to direct experiences of the Divine. Your "inner vision" (the "mind's eye") can help readers in this, and in many other ways. The inner vision is also the seat of many of the intuitive faculties, which are experiencable facts, not imaginings. That means the information obtained by the intuitive faculty is verifiable and reproducibly observable.
>
> To do that, the Balanced Brain is the most efficacious way to function, as well as the most efficient, and the most comfortable. To obtain the Balanced Brain, the person usually needs to spend a great deal of their spare time being receptive, being the "receiver", being accepting and exploring, and not using the analytical intellect, but instead, spending time in the Now and in the Senses and Sensitivities. This is best enjoyed in Natural settings.

Therefore, if we ask: How can we rectify the problem of overemphasizing rationality in information systems, computer engineering and beyond? From Neutrosophic Logic viewpoint, this article recommends that a combination of both the intuitive aspect of the right hemisphere and the analytic or logical thinking processes of the left brain is needed to create a holistic approach. That is why we suggested the term: intuilytics, to capture the essence of the Balanced Brain [3].

7.3 Why Shall We Focus on Data Centers? [5]

As we pointed out earlier, data centers use about 2% of the electricity in the United States; a typical data center has 100 to 200 times the energy use intensity of a commercial building. Data centers present tremendous opportunities—energy use can be reduced as much as 80% between inefficient and efficient data centers (DOE 2011). Data center efficiency measures generally fall into the following categories:

- Power infrastructure (e.g., more efficient uninterruptible power supplies [UPS], power distribution units [PDUs])
- Cooling (e.g., free cooling, variable-speed drives [VSDs], temperature and humidity set points)
- Airflow management (e.g., hot aisle/cold aisle, containment, grommets)
- Information technology (IT).

Grid Scale battery storage

First of all, allow us to mention a trend in data centers: the increasing use of grid scale battery storages. Battery storage is an innovation that empowers power framework administrators and utilities to store energy for some time in the future, hence these reserved energy can be used later on at peak times.

A battery energy stockpiling framework (BESS) is an electrochemical gadget that charges (or gathers energy) from the framework or a force plant and afterward releases that energy sometime in the future to give power or other lattice administrations when required. A few battery sciences are accessible or being scrutinized for framework scale applications, counting lithium-particle, lead-corrosive, redox stream, and liquid salt (counting sodium-based sciences).

See Fig. 7.1.

7.4 A Few Ideas of Energy Efficient Systems

A few ideas for energy efficient P2P network

It is known that most information systems were designed as client–server architecture, where most processes of computation are held in servers (data centers), but that would increase concentration of data and reduce privacy of users. Therefore, there are systems which focus on giving more freedom to users, such as P2P network (peer-to-peer).

One of primary uses of P2P network is for file sharing (like music, video etc.), but that induces larger energy usage, albeit the energy use is not concentrated anymore, but distributed among large base of users (PC, mobile phones etc.). This requires a more complicated method to reduce energy use in such P2P network. After reading some literature on this subject, we come to idea that one of the best method to do so

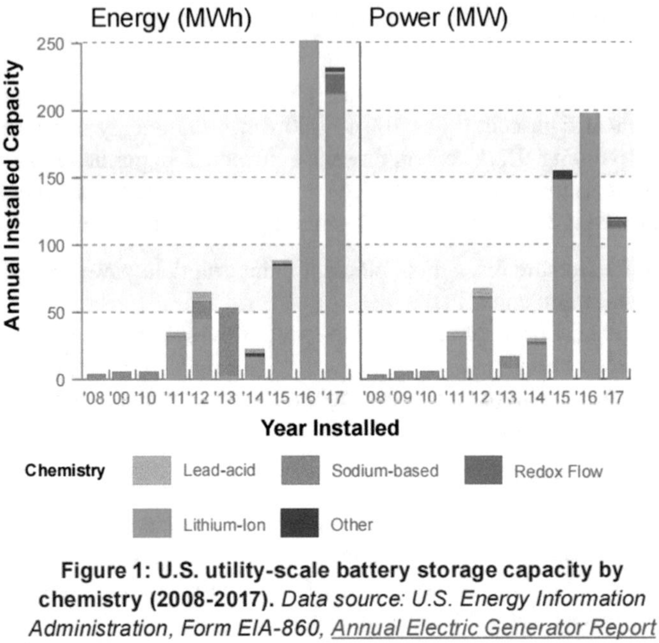

Fig. 7.1 The increasing use of battery storage (*Source* www.greeningthegrid.org)

is by using TCP2P (topological control). In this system, it can be expected that the energy usage in P2P network can be reduced to up to 40–60% [6].

More methods of reducing energy use of P2P network are also discussed in several literatures.

Some suggestions for data centers in four season countries: cool runners

As we all know, four season countries involve quite high temperature at certain seasons and low temperature at winter. For instance, temperature can go down to less than −18 °C in Moscow, less than −20 °C at certain Scandinavian countries, and even less than −40 °C at Tomsk and certain area in Russia Federation.

There are mainly two methods to handle cooling of servers at data centers: air cooling method and direct liquid method (Fig. 7.2).

But there are more radical experiments, one of such radical approach is to expose the data center to outdoor weather (during winter). While this method will clearly lead to reduced energy consumption, several precautions should be used.

A more radical experiment has been performed in Finland, that is to see if servers can work under −20 °C, and it seems the result is quite positive.

Some suggestions for data centers in two seasons/tropical countries

While it is possible to expose the data centers and servers to outdoor weather to reduce energy usage, such a method is not possible to be implemented at tropical countries.

Fig. 7.2 Cooling servers with mineral oil. *Source* Rich Miller, NSA Exploring Use of Mineral Oil to Cool its Servers. Data Center Knowledge. News and analysis for the data center industry

But if we can re-definc the meaning of phrase "energy efficient" to become "reduced energy burden to the grid," then we can think of several ideas to reduce energy burden caused by data centers.

One way is to accelerate implementation of renewable energy (solar and wind) in such data centers, especially in tune with air conditioning methods. Such RE-based air conditioning technologies are already available, or at least they have been reported to be possible. As an example, allow us to remark that an airport located at Banyuwangi city (east side part of East Java, Indonesia) was designed as such to meet physics principles of tropical building especially to avoid air conditioning systems as far as possible, and as a result the airport requires less electrical energy. Such a practice reduces not only monthly electricity bills to the airport operator, but also reduces electricity burden to the national grid. If only such tropical building design principles can be applied to other large airport/facilities, they can lead to massive amount electricity reduction caused by EE (energy efficiency) methods (cf. [4]. Also: [1]).

Table 7.1 depicts how data centers by large platforms have begun to use Renewable energy [2].

But one problem with such a RE-based AC technologies are their intermittency, therefore they need to be connected to the grid.

Is there possible way to reduce such a dependence to the grid? Yes, if we would like to consider non-conventional battery systems, because most batteries available in the market come with quite heavy price tag.

In this regard, allow us to present a summary of our recently published paper on how we can use salt-water electrolyte battery. (Post-script note: For a more recent paper, the readers are advised to check our paper : V. Christianto & F. Smarandache, BPAS Phys. Vol. 41D No. 1, Jun. 2022, url: www.bpasjournals.com)

Summary of our experimental investigation [7]

The effective use of electricity from renewable sources requires large-scale stationary electrical energy storage systems (EESS) with rechargeable high-energy–density, cheap batteries. While batteries using lithium, cadmium, lead-acid etc. have been widely used, there is an alternative source i.e. salt-water which is quite abundant in nature and known as electrolyte. Despite several existing studies on possibilities to implement salt-water battery systems, few seem to have been done in the context of investigating its potential use for low-budget grid-scale electric storage (EESS). One of possible materials for EESS is using electro-chemicals. Electro-chemical energy-storage technologies such as batteries and fuel cell are a method used to store electricity under a chemical form. The technologies differentiated by their energy storage mechanism. Batteries stores energy through electron transfer reaction, wherein the fuel cells store energy from external reactants fed to the cell. This storage technique benefits from the fact that both electrical and chemical energy share the same carrier, the electron.

These preliminary experiments focus on searching of the most effective metal for cathode–anode to generate voltage in salt-water liquid. Experiments also try to find out the most optimum mixture of salt-water.

Electrolyte has a role to conduct ions between the anode and the cathode. Reversible chemical reactions at the electrodes are a vital part of any rechargeable cell. In a battery system there is a redox reaction, where reduction happens on the cathode and oxidation happens on the anode. Those reaction causes the movement of electrons which generate electricity (induced voltage) on a battery. Figure 7.3 illustrates a battery system [7].

$NaCl$ (aqua) $\rightarrow Na^{+} + Cl^{-}$

Zn as cathode and Cu as anode

Cathode: $2H_2O + 2e^{-} \rightarrow 2OH^{-} + H_2$

Anode: $Cu \rightarrow Cu_2{}^{+} + 2e^{-}$

Reaction: Cu(solid) $+ 2\ H_2O$ (liquid) $\rightarrow Cu(OH)_2$ (aqua) $+ H_2$(gas)

In this saltwater experiment, the first material are the copper plate serves as anode and aluminium plate serves as cathode. The second material are the copper (Cu)

Table 7.1 Data centers powered by 100% renewable energy

Data center company	Facility location	Power capacity (MW)	% Resource mix of local utility (natural gas–nuclear–coal)	PUE
Google	Council Bluffs, IA	105	19%–6%–45%	1.12 (average)
	Pryor, OK	49	25%–0%–46%	
	Hamina, Finland	19	13%–32%–22%	
Apple	Prineville, OR	2	12%–0%–60%	1.5 (North Carolina)
	Maiden, NC	19	4%–57%–38%	
	Newark, CA	15	27%–21%–0%	
	Reno, NV	5	51%–15%–7%	
Facebook	Lulea, Sweden	70	1%–40%–1%	1.07 (Lulea, Sweden)
	Altoona, IA	70	19%–6%–45%	
Microsoft	San Antonio, TX	27	37%–13%–27%	1.125
Yahoo	Lockport, NY	23	36%–23%–12%	1.08

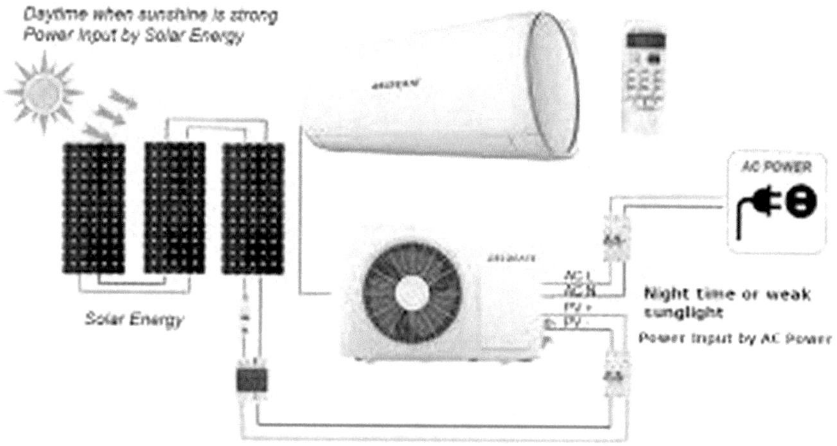

Fig. 7.3 Solar-powered air conditioning system

plate serves as anode and zinc plate serves as cathode to get the better current flow (Fig. 7.4).

While our experiment is still in preliminary phase, we can expect that such method can be developed further, therefore data centers's cooling system can be built and supported by renewable energy powered air conditioning, which in turn it can be designed as such to reduce dependence to national electric grid. See also [8–10] (Fig. 7.5).

On trend of chemical energy storage

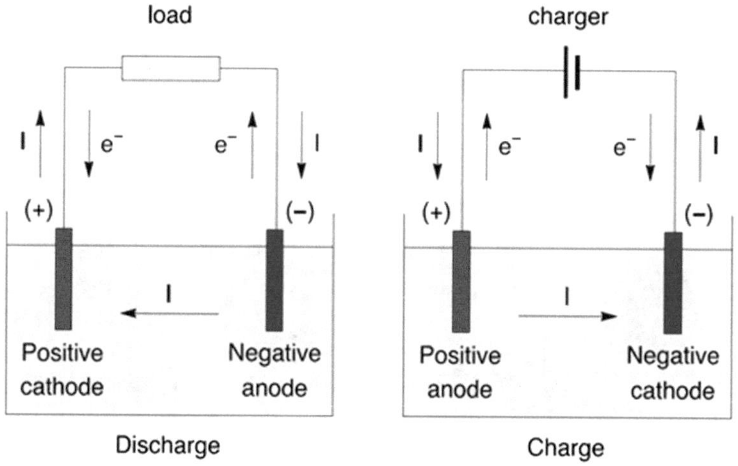

Fig. 7.4 Illustration of a battery system. *Source* Illustration of a battery system. Download scientific diagram (researchgate.net)

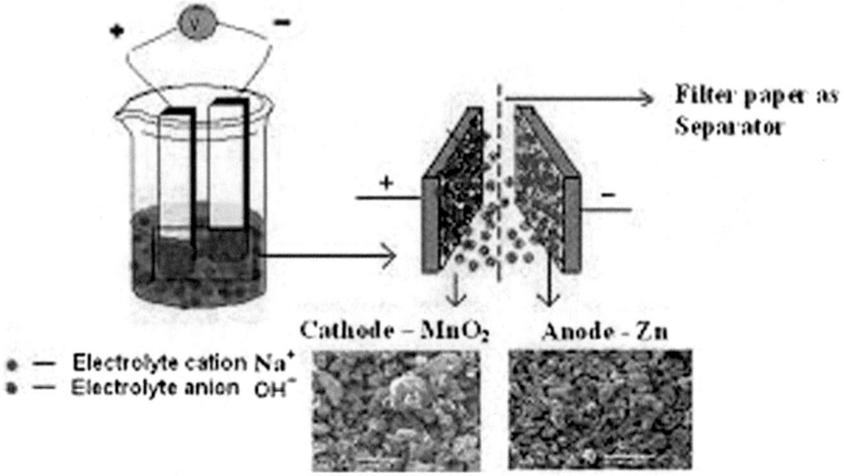

Fig. 7.5 An example of salt-water battery system. *Source* Electronic products. How salt water batteries can be used for safe, clean energy storage—Electronic Products

Huge scope battery stockpiling frameworks are progressively being utilized across the force matrix in the United States. In 2010, 7 battery stockpiling frameworks represented just 59 megawatts (MW) of force limit, the most extreme measure of force yield a battery can give in any moment, in the US. By 2015, 49 frameworks represented 351 MW of force limit. This development proceeded at an expanded rate for the following three years, and the absolute number of operational battery stockpiling frameworks has more than multiplied to 125 for a sum of 869 MW of introduced power limit as of the finish of 2018.

The most punctual enormous scope battery stockpiling establishments in the US utilized nickel-based and sodium based sciences (Fig. 7.6). Notwithstanding, since 2011, most establishments have picked lithium-particle batteries, including retrofits

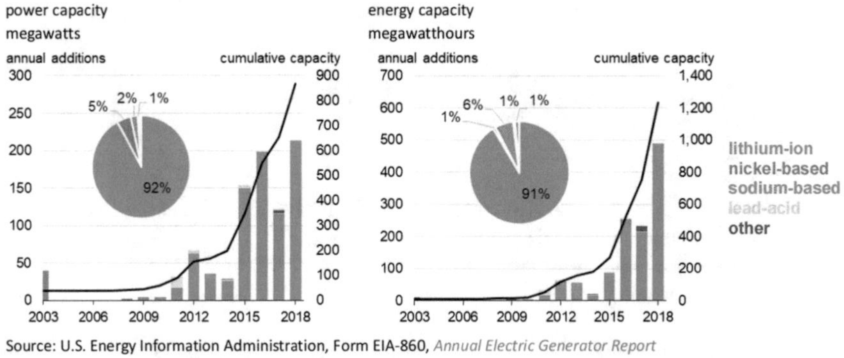

Fig. 7.6 Large-scale battery storage capacity by chemistry (2003–2018)

Fig. 7.7 Switzerland's largest salt-water battery. *Source* Switzerland's largest saltwater battery storage in operation—BlueSky Energy. Stromspeicher Batterien (bluesky-energy.eu)

of more seasoned frameworks that at first depended on various sciences. For instance, in 2012, Duke Energy added a day and a half of lead-corrosive battery stockpiling to its Notrees wind power office in West Texas. At the point when the lead-corrosive batteries were first introduced, the battery framework partook in the area's recurrence guideline market, which required fast charging and releasing that essentially corrupted the batteries. In 2016, Duke Energy supplanted the first lead-corrosive batteries with better performing lithium-particle batteries.[1]

Generally speaking, there is increasing demand for electrochemical batteries of various types. The following figure tells more, although it does not tell on the coming age of salt-water battery storage.

As of salt-water battery in real use, there is largest salt-water battery being used in Switzerland (Fig. 7.7).

Small scale energy storage trends

Other than grid scale energy storage, there is also trend of small scale energy storage.

In 2018, utilities revealed 234 MW of existing limited scope stockpiling power limit in the US. A minimal over half of this limit was introduced in the business area, 31% was introduced in the private area, and 15% was introduced in the modern area. The excess 3% was straightforwardly associated with the dissemination lattice, for example, by the utility at their own conveyance substation.

The information gathered for limited scope applications rely upon the electric utility's admittance to data about establishments in its domain. In the event that end clients of capacity frameworks are introducing frameworks for purposes where the

[1] US Department of Energy. Battery Storage in the United States: An Update on Market Trends. Independent Statistics and Analysis. July 2020. www.eia.gov.

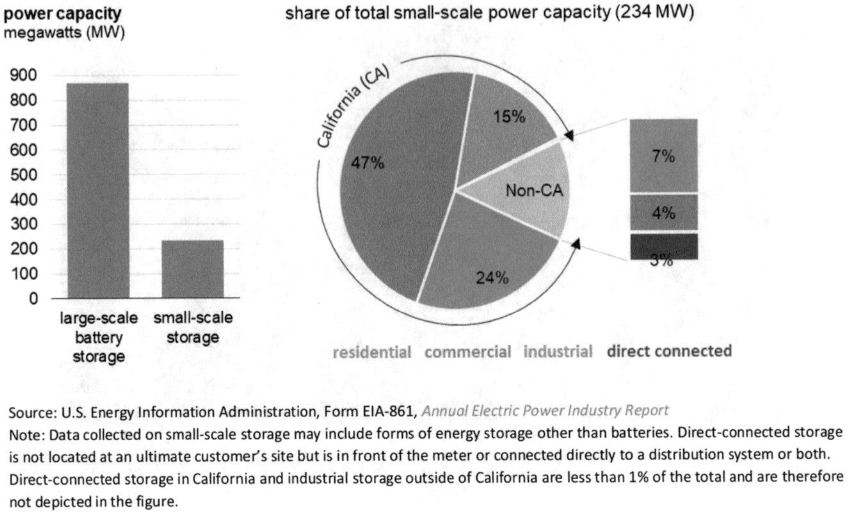
power capacity
megawatts (MW)
900
800
700
600
500
400
300
200
100
0
large-scale battery storage
small-scale storage
share of total small-scale power capacity (234 MW)
California (CA)
47%
15%
Non-CA
24%
7%
4%
3%
residential commercial industrial direct connected
Source: U.S. Energy Information Administration, Form EIA-861, Annual Electric Power Industry Report
Note: Data collected on small-scale storage may include forms of energy storage other than batteries. Direct-connected storage is not located at an ultimate customer's site but is in front of the meter or connected directly to a distribution system or both. Direct-connected storage in California and industrial storage outside of California are less than 1% of the total and are therefore not depicted in the figure.

Fig. 7.8 Small scale energy storage trend in California (USA, 2018)

framework would not associate with the dispersion organization—for instance back-up applications—the power appropriation utility may not think about those framework establishments. Utilities gather data on limited scope stockpiling frameworks basically through between association arrangements. Since these arrangements are planned by the utilities, the data about capacity units may not be gathered in a steady organization across all utilities.

As shown in Fig. 7.8, in 2018, 86% of revealed limited scope stockpiling power limit in the US was in California and, explicitly, was possessed by six utilities: Southern California Edison (SCE), Pacific Gas also, Electric (PGE), San Diego Gas and Electric (SDGE), Los Angeles Division of Water and Force (LADWP), Sacramento Civil Utility Area (SMUD), and City of Moreno Valley. In 2018, most establishments of limited scope stockpiling in the business area in California were in SCE's domain (64% of such limit) and SDGE's domain (22%). Most establishments (95%) of limited scope stockpiling in the mechanical area in California were in PGE's domain.

7.5 Conclusions

In this paper, we present some recent data from available sources, how data centers tend to become more environmentally friendly, and this can be a good trend for the future.

While some the aforementioned ideas may sound a bit elementary, we have presented a number of ideas from physicist perspective, suggesting that significant energy reduction and conservation can be achieved in foreseeable future. For more references, the readers are referred to [11–13].

All in all, some of these ideas came merely from physicist's perspective.

Version 1.0: 29 March 2021, pk. 15:11.

Version 1.1: 29 March 2021, pk. 16:42.

Version 1.2: 08 June 2021, pk. 18:55.

FS, VC, YU.

References

1. Bulbaai, R., Halman, J.I.M.: Energy efficient building design for tropical climate. Sustainability **13**, 13274 (2021). https://doi.org/10.3390/su132313274
2. Chalise, S., et al.: Data Center Energy Systems: Current Technology and Future Direction. IEEE (2015)
3. Christianto, V., Boyd, R.N., Smarandache, F.: How to balance the intuitive and analytical function of the human brain. EC Neurol. **11**(7), 495–499 (2019)
4. Christian, P.: Sintesis arsitektur lokal dan modern pada atap Bandara Banyuwangi. Undergraduate Thesis, Fakultas Teknik, Univ. Parahiyangan, West Java (2018)
5. Huang, R., Masanet, E.: Data center IT efficiency measures. Chapter 20 in The Uniform Methods Project: Methods for Determining Energy Efficiency Savings for Specific Measures. NREL/SR-7A40-63181, January (2015)
6. Nurminen, J., Noyranen, J.: Energy-consumption in mobile peer-to-peer—quantitative results from file sharing. Paper Peer Reviewed at the Direction of IEEE Communications Society Subject Matter Experts for Publication in the IEEE CCNC 2008 Proceedings (2008)
7. Umniyati, Y., Indra Wijaya, K., Sinaga, E., Christianto, V.: Preliminary experiments on potential use of salt-water battery for cheap electric storage: work in progress. Jurnal Ilmu dan Inovasi Fisika **5**(1), 74–81 (2021)
8. Moia, D., Giovannitti, A., Szumska, A.A., et al.: A salt water battery with high stability and charging rates made from solution processed conjugated polymers with polar side chains (2017). arXiv: [1711.10457]
9. Smarandache, F., Christianto, V.: Observation of anomalous potential electric energy in distilled water under solar heating (2010). Vixra: 1003.0090
10. Pollack, G.W.: Batteries made from water. In: Proceedings of the NPA. www.billhowell.ca
11. Muralidhar, R., et al.: Energy efficient computing systems: architectures, abstractions and modeling to techniques and standards, 22 July 2020. arXiv:2007.09976v2 [cs.AR]
12. Tand, L., et al.: Energy efficient and reliable routing algorithm for wireless sensors networks. Appl. Sci. **10**, 1885 (2020). https://doi.org/10.3390/app10051885
13. US Department of Energy: Battery storage in the United States: an update on market trends. Independent statistics and analysis, July 2020. www.eia.gov

Chapter 8
The Dark Side of Technology Use: The Relationship Between Technostress Creators, Employee Work-Life Balance, and Job Burnout While Working Remotely During the COVID-19 Lockdown

Živilė Stankevičiūtė

Abstract Rapid progress of digital technologies and their increasing spreading into the work domain have been the subject of turbulent discussions for the last decade. The adaptation and use of information and communication technologies (ICTs) led to redefinition of organisational structures and the way employees work, making it possible to connect anytime, anywhere and deliver data in real time. Recently, the use of ICTs for working purposes has increased due to the COVID-19 pandemic. The spread of COVID-19 has been changing working habits around the world with employers encouraging or even insisting that employees work remotely. As such, work is mainly based on ICT use while experiencing some technology-related job demands, which are generally named techno-stressors or technostress creators. Previous research provides some evidence that the ubiquity of technologies can add to employees experiencing technostress because of the increased workload, excessive technology dependence, demands for enhanced productivity, and a constant need to adapt to emerging ICT applications, functionalities, and workflows. There is still a gap in the literature about the impact of technology use on employee well-being while working remotely during the COVID-19 lockdown. This chapter aims at revealing the relationship between the construct of technostress creators, including its five dimensions, and employee well-being in terms of work-life balance and job burnout. In doing this, quantitative data were collected in Lithuania. The results support the idea that technostress creators have a negative effect on employee well-being. Moreover, the findings suggest that techno-overload and techno-invasion reduce work-life balance and lead to job burnout. The chapter has strong practical implications seeing that the results are in line with the idea that organisations should create and implement policies and practices for reducing the impact of technostress creators.

Ž. Stankevičiūtė (✉)
School of Economics and Business, Kaunas University of Technology, Gedimino g. 50, 44249 Kaunas, Lithuania
e-mail: zivile.stankeviciute@ktu.lt

L. Ivascu et al. (eds.), *Intelligent Techniques for Efficient Use of Valuable Resources*, Intelligent Systems Reference Library 227,
https://doi.org/10.1007/978-3-031-09928-1_8

Keywords Technostress · ICTs · Technostress creators · Work-life balance · Job burnout · Employee well-being

8.1 Introduction

Over the centuries, work has been a subject of transformation [59]. Turning to the last decades, particularly, rapid advances in technology have dramatically influenced organisations and people [22, 23, 69, 40, 44, 45, 51]. ICTs have become crucial components of working environments and serve as important working tools [37]. While addressing organisational structures, division of work, and relationships between people, ICTs affect knowledge-intensive work as well as manual and industrial work [37]. Recently, the spread of COVID-19 has been accelerating the use of ICTs worldwide as employees were encouraged or even forced to work remotely [45, 61]. Work from home supported by technologies has become a rule that particularly promoted technostress, which is defined as "stress experienced by end users in organizations as a result of their use of ICTs" [49, pp. 417–418].

Prior literature on technostress has generally focused on several areas. First, the conceptualisation of technostress and the factors that cause stress, called technostress creators, are the focus of interest [45, 66]. Second, the antecedents of technostress [4, 60, 62] have been examined. Third, researchers have been keen to study the consequences of technostress in terms of employee attitudes, health, well-being in general, and performance or organisational outcomes [18, 49, 64]. The current book chapter addresses two specific streams, namely, the technostress creators and the consequences of technostress.

Despite the numerous calls for research on the topic, several gaps could still be noted. First, the majority of previous studies described the technostress creators as a multidimensional superordinate construct that includes five dimensions: techno-overload, techno-invasion, techno-complexity, techno-insecurity, and techno-uncertainty [66]. The individual impact that each dimension has on employee or organisational outcomes has not yet been sufficiently examined [18]. Trying to narrow this gap, this chapter explores not only the impact the technostress creators have on employees overall, but also the impact caused by a particular dimension. Second, regarding the outcomes of technostress creators for employees, the prior literature included mainly job satisfaction [49, 67], organisational commitment [2, 49, 67], and work-family conflict [45], whereas work-life balance [52] or job burnout [36, 75] received less attention. Accordingly, the chapter addresses the mentioned two constructs of employee well-being while adding empirical evidence how technostress creators effect work-life balance and job burnout.

The aim of the chapter is to reveal the linkage between the technostress creators in general and the five individual dimensions comprising the construct, namely techno-overload, techno-invasion, techno-complexity, techno-insecurity, and techno-uncertainty, to work-life balance and job burnout while working remotely during the COVID-19 pandemic. In doing this, the chapter seeks to answer the following: (a) whether and to what extent do the employees experience technostress creators in general and each separate technostress creator? (b) whether and to what extent do the employees feel work-life balance and job burnout? (c) will and how technostress creators, including their five dimensions, impact the work-life balance and job burnout of employees? To answer these questions, this chapter analyses data from the survey carried out in Lithuania, a country known for world 's fastest consumer download speed and second fastest upload speed [33] creating excellent conditions for using the ICTs. The sample consisted of employees working only remotely during the COVID-19 lockdown.

The chapter intends to contribute to the literature in several ways. First, the intention is to respond to the recent call of Chandra et al. [18] and not only address technostress creators as a multidimensional construct, but also analyse each dimension of the construct. As five dimensions represent different facets of the general construct, the deconstructed way helps understand the phenomenon more deeply [18]. Next, the chapter addresses a specific context and sample. As the research was conducted during a crisis situation (COVID-19 pandemic) while employees were forced to work from homes, the chapter contributes to the body of literature dealing with the ICTs' impact on employee well-being under unpredictable conditions and extreme events, without any prior experience how to deal with them. Finally, in order to understand how technostress creators are related to work-life balance and job burnout, a Job Demand-Resource (JD-R) model [21] and border theory [19] were applied.

The remainder of the chapter is structured as follows. The theoretical part gives an overview of the literature on changing work nature as a result of ICTs and remote work, and describes technostress and its creators. Later, the hypotheses are developed referring to particular theories. Then, the research method applied is described. The empirical results and discussion come further. Finally, conclusions are drawn.

8.2 Changing Work Nature as a Result of ICTs and Remote Working

From the scope of this chapter, two main factors influencing and redesigning work nature are extremely relevant, namely ICTs and remote working.

Lately, new technologies and recent advances in telecommunication have been among the factors that changed work substantially [44]. The adaptation and use of ICTs have led to rethinking the manner in which businesses create and capture value [58], to redefinition of organisational structures [49], and to reconsidering where, when and how individuals do their work and the ways in which employees interact [28]. ICTs refer to a combination of hardware, software, and communication networks [12] that enable capture, storing, processing, and transfer of electronic information [51]. Organisations can easily provide employees with plenty of technological tools making work a major field of digitalisation [24]. Although the use of ICTs pays back for the organisations in terms of reduced operational costs, greater process efficiency, new strategic alternatives, or possibilities for innovations [66], the discussion about the "dark side" of technologies [47, 53, 54] or "dichotomy" [67] is not new. Naturally, the notion of not only positive but also the detrimental effects of ICTs, especially on the employees and their well-being, has been drawing interest since computers and computerised technology were widely introduced in the 1980s [47]. Referring to dichotomy, ICTs enable employees to "quickly and easily access information, work from anywhere, and share information and insights with colleagues in real time" [67, p. 114]. However, workflow applications, communication devices and other technologies can make employees "feel compulsive about being connected, forced to respond to work-related information in real time, trapped in almost habitual multitasking and left with little time to spend on sustained thinking and creative analysis" [67, p. 114].

In general, three main characteristics describe the interaction between ICTs and work environment [49, 65]. First, as ICTs frequently require employees to process information simultaneously and continually from different devices, the people deal with a surfeit of information, experience frequent interruptions from different ICTs, and engage in multitasking [65]. Distraction by incoming information and constant interruption refers to a state known as "continuous partial attention" [31]. Second, because of the constantly released new versions of ICTs, there is a significant gap between the employees' current knowledge and the knowledge they need to master the ICTs [49]. Third, ICTs have the potential to change the work culture with increased possibilities for remote supervision or social isolation [49]. Drawing upon the mentioned characteristics and the notion that employees become increasingly dependent on ICTs [62], digital technologies may be perceived as stressful, and accordingly such stress is labelled as technostress [24].

Another factor promoting changes in the nature of work is remote working. Remote working, also known as teleworking, is an arrangement between the employee and employer where the employee performs work remotely outside the employer's premises aided by ICTs [26]. In the early days, since the 1970s, as ICTs started increasingly permeate professional and private life, it was expected that, at some point in the future, everyone would work remotely [26]. However, the adoption of teleworking progressed much slower than anticipated [72]. The arguments behind this were various human, social and organisational factors, including fundamental constraints associated with the individual's need to meet other people face-to-face [70]. However, the coronavirus disease and the lockdown as the means to prevent the spread of COVID-19 accelerated the process of remote working implementation. Remote working has become a widespread solution [45] with its advantages and disadvantages. As regards the positive side of remote working, it causes improved performance, cutting the costs of traveling to-from office, saving time and organisational resources, and increasing employee satisfaction [8, 68]. This notwithstanding, the other side of the coin is described as generating some negative consequences particularly in relation to employee well-being, and it can cause stress, discomfort, and anxiety also due to the constant use of ICTs [61]. In this context, the focus is again placed on technostress, which is caused by the employee's attempts to follow the requirements and to use ICTs for achieving performance results.

8.3 Technostress and Technostress Creators

According to McGrath [43], stress is defined as a state experienced by an individual when there is an "environmental situation that is perceived as presenting a demand which threatens to exceed the person's capabilities and resources for meeting it, under conditions where he or she expects a substantial differential in the rewards and costs from meeting the demand versus not meeting it" [43, p. 1351]. The Transaction-Based Approach [38, 43] addresses three components, namely stressors, strain, and situational variables, as the foundation to study stress. *Stressors* represent events, demands, stimuli or conditions that create stress; situational *variables* are organisational mechanisms that can buffer or reduce the impact of stressors; while *strain* refers to the behavioural, psychological, and physiological outcomes of stress that are observed in individuals [49].

The term technostress was originally coined in 1984 by Craig Brod [15], who described it as a disease caused by one's inability to cope or deal with ICTs in a healthy manner. More recently, Tarafdar et al. [65] referred to technostress as "the stress that users experience as a result of application multitasking, constant connectivity, information overload, frequent system upgrades and consequent uncertainty, continual relearning and consequent job-related insecurities, and technical problems associated with the organizational use of ICT" [pp. 304–305]. One of the widely accepted definitions nevertheless belongs to Ragu-Nathan et al. [49] claiming that "technostress relates to the phenomenon of stress experienced by end users in organizations as a result of their use of ICTs" [pp. 417–418].

Technology-related job demands that can provoke technostress are generally named techno-stressors or technostress creators [66]. Tarafdar et al. [66] proposed a classification, widely accepted in the literature, of five technostress creators: techno-overload, techno-invasion, techno-complexity, techno-insecurity, and techno-uncertainty.

Techno-overload describes situations where use of ICTs forces employees to work faster and longer [65]. ICTs make it possible to process simultaneous streams of real-time information, resulting in information overload, interruptions, and multitasking. Regarding information overload, employees get more information than they can efficiently handle and use [67], leading to "information fatigue" [71] and "data smog" [13]. Interruptions, for example email alerts, force the employees to look at information as soon as it arrives, making it difficult to sustain mental attention [67]. Finally, the aim of multitasking is to try to do more in less time. However, there are some limits and the use of ICTs pushes employees to exceed these limits [67].

Techno-invasion refers to being "always exposed" so that employees can be reached almost anywhere and anytime [62]. Employees feel the need to be constantly connected, workdays extend into family hours, it is almost impossible to "cut away" and, finally, not connecting becomes disquieting [67].

Techno-complexity refers to the situations where ICTs' complexity and features make employees feel inadequate about their skills [65]. Employees are forced to spend much time and effort learning how to use new ICTs and it seems that ICTs' producers or vendors are not willing to facilitate the learning as the manuals can be unwieldy and full of jargon [49, 67].

Techno-insecurity is associated with the employee feeling of being threatened about losing their jobs due to replacement by robots, machines, AI or other people who have better knowledge of ICT [65].

Techno-uncertainty is related to short life cycles of ICTs. Continuous changes and upgrades do not offer the employees a chance to develop a base of experience for a particular system [67]. Their knowledge becomes rapidly outdated, and they are required to re-learn things often and at high pace [49, 65].

In summary, ICTs create stress because they generate information overload, create interruptions, lead to multitasking, change frequently, require to invest huge amounts of efforts and time into upgrading skills, and never let feel "free" of them (always connected).

8.4 Linkage Between Technostress Creators and Work-Life Balance: Hypothesis Development

In recent years, a great deal of emphasis has been placed on the employee well-being [30], and as such, work-life balance has been receiving growing attention [73] with the intention to support the employees in reconciling the competing demands of their paid work and non-work responsibilities [1]. Work-life balance is defined as "the relationship between work and non-work aspects of individuals' lives, where achieving a satisfactory work-life balance is normally understood as restricting one side (usually work), to have more time for the other" [35]. Such definition raises the question and encourages debates of what constitutes "balance". Some scholars perceive balance as inferring an equal distribution of energy, time, and commitment to work and non-work roles [29], while others adopt a "situationist" approach [50], where balance depends on the individual's circumstances.

At the initial stage of work-life balance conceptualisation, work and non-work (family) roles were clearly separated in terms of time and space, arguing that work occurs during designated hours and at a place away from home [1]. However, ICTs have been changing the dominant understanding, as work can now be done at any time and anywhere [67]. Moreover, remote working during the COVID-19 lockdown has made working at home even compulsory [32, 48].

Work/family border theory examines the way people navigate the work and non-work domains and manage the borders between them in order to strike a balance [19]. In this chapter, border theory is invoked to examine the impact of ICTs on the border between work and life domains. According to Clark [19, p. 756], borders are "lines of demarcation between domains defining the point at which domain-relevant behavior begins or ends". There are three main types of borders: temporal, physical, and psychological [19]. The temporal and physical borders define when and where domain-relevant behaviours take place, usually in terms of time and space (defined working hours or walls, and doors). The psychological borders are largely self-created rules that indicate different thinking patterns, emotions, and behaviours [19].

Actually, the transition from one domain to another is not always easy, especially where the borders between home and work are intentionally blurred as is the case for remote workers [27]. Accordingly, the prediction of border theory is that remote working will heighten negative work-home spillover [27, 39, 41].

Turning to the construct of technostress creators and its dimensions, it seems that work-life balance is significantly affected by ICTs [1, 45].

Regarding techno-overload, the situations of "too much" where employees are being forced to do more and faster necessitate working longer hours and this does not support the work-life balance as more time is required for work activities. The "always connected" approach (techno-invasion) makes it even impossible to take vacations and this does not go hand in hand with work-life balance. Accordingly, difficulties with ICTs (techno-complexity), insecurity about the job (techno-insecurity) and constant changes (techno-uncertainty) make the line between work and life even more blurred where work requires more employee involvement. Accordingly, the chapter hypotheses the following:

H1. Technostress creators will be negatively related to work-life balance.

H1a. Techno-overload will be negatively related to work-life balance.

H1b. Techno-invasion will be negatively related to work-life balance.

H1c. Techno-complexity will be negatively related to work-life balance.

H1d. Techno-insecurity will be negatively related to work-life balance.

H1e. Techno-uncertainty will be negatively related to work-life balance.

8.5 Linkage Between Technostress Creators and Job Burnout: Hypothesis Development

JD-R model was initially introduced to understand burnout [5, 21] and later it was supplemented with work engagement [7, 57]. At the heart of the JD-R model [5, 6, 21] lies the assumption that risk factors associated with job stress can be classified in two general categories (i.e. job demands and job resources) and accordingly employee health and well-being is an outcome of balance between job demand and resources. According to Demerouti et al. [21, p. 501], job demands are "physical, social and organizational aspects of the job that require sustained physical or mental effort and are therefore associated with certain physiological and psychological costs". Thus, job demands refer to "bad things" at work that drain energy [57], such as work overload, future job insecurity, or time pressure. In contrary, job resources are perceived as a "good things" referring to "physical, psychological, social or organizational aspects of the job that may do any of the following: (a) be functional in achieving work goals; (b) reduce job demands and the associated physiological and psychological costs; (c) stimulate personal growth and development" [21, p. 501]. Examples of job resources are support from others and performance feedback [57].

As it was mentioned before, the JD-R model was introduced to understand burnout. Job burnout refers to "a chronic state of work-related psychological stress that is characterised by exhaustion (i.e. feeling emotionally drained and used up), mental distancing (i.e. cynicism and lack of enthusiasm), and reduced personal efficacy (i.e. doubting about one's competence and contribution at work)" [57, pp. 120–121].

Exhaustion refers to feelings of being overextended and depleted [42]. Cynicism represents indifferent or distant attitude towards work in general, and work colleagues, while work is not seen as meaningful because the interest in work has been lost [55]. Reduced professional efficacy refers to feelings of incompetence and a lack of successful achievement and productivity at work [42].

Turning to ICTs, it could be stated that technostress creators in general and the five specific dimensions, namely techno-overload, techno-invasion, techno-complexity, techno-insecurity, and techno-uncertainty, serve as job demands, which drain energy [57] and enhance job burnout. For instance, the study of teleworkers by Sardeshmukh et al. [56] revealed that time pressure, role ambiguity, and role conflict led to exhaustion.

In case of techno-overload, it becomes difficult for employees to set priorities regarding new information as the information overload and multitasking are prevailing [65]. Previous studies have demonstrated that feeling of information overload is associated with the loss of control over the situation while being overwhelmed and this in turn may cause negative effect on employee well-being [9]. Due to techno-invasion, privacy is lost [65] and this leads to work-home conflict, which in turn results in work exhaustion [3]. Techno-complexity requires spending more energy and time to updating skills in the use of ICTs, role conflict and role overload as job demands manifest and, based on prior research, lead to burnout among employees [34, 74]. Techno-insecurity is related to job insecurity [20], and accordingly has a huge potential to increase job burnout [59]. Finally, techno-uncertainty with continuous changes or upgrades in ICTs may generate the lack of the control (job demand) which may lead to higher burnout. Accordingly, the chapter hypotheses the following:

H2. Technostress creators will be positively related to job burnout.

H2a. Techno-overload will be positively related to job burnout.

H2b. Techno-invasion will be positively related to job burnout.

H2c. Techno-complexity will be positively related to job burnout.

H2d. Techno-insecurity will be positively related to job burnout.

H2e. Techno-uncertainty will be positively related to job burnout.

8.6 Methodology

Sample and data collection. Given the objective of the research, data were collected by using a convenience sampling type from employees in Lithuania working only remotely during the COVID-19 lockdown. Convenience sampling is a type of non-probability sampling where members of the target population who meet certain practical criteria, such as availability at a given time, easy accessibility, geographical proximity, or the willingness to participate are included for the purpose of the study [25].

For the survey, an online questionnaire was created. The questionnaire was distributed via LinkedIn, Facebook and other social networks. Due to the way of questionnaire dissemination, it is impossible to estimate the number of persons the questionnaire was sent to and the response rate. While distributing the questionnaire, the information about the purpose of the survey and a link to a survey were enclosed. The first question was related to the nature of work. The possibility to continue with the survey was provided only in case of answering that they worked remotely full-time during the lockdown. Data collection took place during the COVID-19 lockdown period, in May and June, 2021 (approx. 1 month). At the end of the research, 138 questionnaires were collected. The profile of respondents is presented in Table 8.1. Turning to demographical characteristics of the respondents, 122 of them were women (88.4%). Seventy-two respondents were born in 1982–2000 and 62 respondents were born in 1961–1981. Almost all respondents held a university degree.

Instrument. A self-reported questionnaire with a five-point Likert scale was used in the study where 1 indicated "strongly disagree", and 5 indicated "strongly agree". All items were translated into the Lithuanian language using a back translation procedure [14], ensuring translation accuracy.

Measures. All items regarding technostress creators started with the phrase: "while working remotely during lockdown…". Technostress creators were measured using a 23-item scale provided by Srivastava et al. [62]. The examples of techno-overload items are: "I am forced by ICTs to work much faster" and "I am forced by ICTs to do

Table 8.1 The respondents' profile

Characteristics	Frequency (n)	Percentage (%)
Gender		
Female	122	88.4
Male	16	11.6
Year of birth		
Born in 1982–2000	72	52.2
Born in 1961–1981	62	44.9
Born in 1943–1960	4	2.9
Education level		
University degree	130	94.2
Higher non-university education	4	2.9
Post-secondary education	4	2.9

more work than I can handle". The examples of techno-invasion items are: "Because of ICTs I spend less time with my family" and "Because of ICTs I have to be in touch with my work even during my vacation". Items like "I do not find enough time to study and upgrade my ICT skills" and "I need a long time to understand and use new ICTs" serve as examples of how techno-complexity was measured. The example of techno-insecurity item is "Because of new ICTs, I feel constant threat to my job security". Finally, the example of techno-uncertainty item is "In our organisation, there are always constant changes in ICT software".

Cronbach's alpha was $\alpha = 0.819$ for techno-overload; $\alpha = 0.844$ for techno-invasion; $\alpha = 0.836$ for techno-complexity; $\alpha = 0.759$ for techno-insecurity; $\alpha = 0.906$ for techno-uncertainty, and $\alpha = 0.872$ for technostress creators as a second-order construct.

Work-life balance was measured on a four-item scale proposed by Brough et al. [16]. The four items were: "I currently have a good balance between the time I spend at work and the time I have available for non-work activities", "I have difficulty balancing my work and non-work activities" (R), "I feel that the balance between my work demands and non-work activities is currently about right" and 'Overall, I believe that my work and non-work life are balanced". Cronbach's alpha was $\alpha = 0.779$.

Finally, job burnout was measured by a 9-item scale proposed by Sarmela-Aro et al. [55], which is based on the Bergen Burnout Inventory (BBI). The examples of items sound as follows: "I often sleep poorly because of the situation at work" and "I feel that I am gradually losing interest in my customers or my other employees". Cronbach's alpha was $\alpha = 0.881$.

Thus, all measures were subjected to reliability analysis. As all Cronbach's alpha coefficients exceeded 0.7, all measures were considered acceptable for the analysis [46].

8.7 Results

The means, standard deviations for the scales and correlation matrix are provided in Table 8.2.

As it is seen from Table 8.2, the mean of ratings of work-life balance was 2.82, while job burnout was rated with a mean of 2.69. According to the respondents, while working remotely, they did not feel a high level of technostress creators (mean = 2.08).

Referring to Table 8.2, a negative correlation between techno-overload and work-life balance (-0.316, $p < 0.01$), between techno-invasion and work-life balance (-0.406, $p < 0.01$), and between technostress creators and work-life balance (-0.205, $p < 0.05$) has been revealed. In case of techno-complexity, techno-insecurity, and techno-uncertainty, no significant relationships with work-life balance have been found.

Table 8.2 Mean, standard deviation and correlations

Title	Mean	SD	1	2	3	4	5	6	7	8	9	10	11
1. Age	2.49	0.557											
2. Gender	1.12	0.321	−0.240**										
3. Education level	5.91	0.372	−0.074	0.085									
4. Techno-overload	3.38	0.880	−0.078	−0.210*	0.102								
5. Techno-invasion	3.01	1.10	−0.276**	−0.220**	0.154	0.453**							
6. Techno-complexity	2.17	0.822	−0.182*	−0.068	0.032	0.318**	0.424**						
7. Techno-insecurity	2.08	0.727	−0.175*	−0.154	0.081	0.334**	0.442**	0.574**					
8. Techno-uncertainty	3.18	0.939	−0.151	0.108	−0.026	0.268**	0.074	−0.062	−0.126				
9. Technostress creators	2.74	0.576	−0.265**	−0.176*	0.110	0.749**	0.757**	0.697**	0.675**	0.343**			
10. Work-life balance	2.82	0.903	0.095	0.059	−0.057	−0.316**	−0.406**	0.079	0.026	0.013	−0.205*		
11. Job burnout	2.69	0.869	−0.073	−0.267**	0.098	0.344**	0.612**	0.463**	0.522**	−0.093	0.579**	−0.461**	

Note n = 138; * $p < 0.05$; ** $p < 0.01$

Concerning job burnout, no significant relationships were found only in case of techno-uncertainty (Table 8.2).

To test the study hypotheses, multiple regression analyses were conducted (Table 8.3).

H1 proposed a negative relationship between technostress creators as a second-order construct and work-life balance. As it is seen from Table 8.3, technostress creators diminish work-life balance (−0.205, $p < 0.05$). Thus, H1 was supported.

Concerning H1a, H1b, H1c, H1d and H1, only H1a and H1b were supported demonstrating that techno-overload (−0.316, $p < 0.01$) and techno-invasion (−0.406, $p < 0.01$) lowered work-life balance of remotely working employees during the COVID-19 lockdown.

According to H2, a positive association was predicted between technostress creators as a second-order construct and job burnout. This hypothesis was also confirmed, since it can be seen from Table 8.3 that technostress creators enhanced job burnout (−0.579, $p < 0.01$). Accordingly, as illustrated in Table 8.3, H2a, H2b, H2c and H2d were supported. In case of H2e, no statistically significant relationship was found.

8.8 Discussion

The chapter was intended to examine the linkage between the construct of technostress creators and its five dimensions (techno-overload, techno-invasion, techno-complexity, techno-insecurity, and techno-uncertainty) and work-life balance and job burnout while working remotely during the COVID-19 lockdown. More specifically, drawing upon the JD-R model and border theory, the chapter claimed for a negative association in case of work-life balance and for a positive association regarding job burnout. In doing this, the chapter echoes the call in the previous literature to focus on the impact ICTs have on employee well-being while working remotely [70]. Moreover, the chapter addressed the specific lockdown aspect seeing that the survey was carried during COVID-19 lockdown period. Further, theoretical and practical implications of the findings are discussed.

Theoretical implications

First, the earlier literature described technostress creators as a multidimensional second-order construct that comprises five dimensions [65], while the individual impact of various outcomes of each dimension was largely ignored [18]. Seeking to "more deeply understand the phenomenon" [65, p. 19], the chapter provides insights and empirical evidence that each of the five technostress creators uniquely (individually) contributes to work-life balance and job burnout.

Second, the hypothesis regarding linkage between technostress creators and two employee well-being outcomes, namely work-life balance and job burnout, was theoretically substantiated by the JD-R model and border theory. Hence, by categorising the technostress creators as job demands [57] and by rethinking temporal and physical

Table 8.3 Regression analysis

	Dependent variable (Standardized)							
	Work-life balance H1	Work-life balance H1a	Work-life balance H1b	Job burnout H2	Job burnout H2a	Job burnout H2b	Job burnout H2c	Job burnout H2d
Independent variables								
Techno-overload		−0.316***			0.344***			
Techno-invasion			−0.406***			0.612***		
Techno-complexity							0.463***	
Techno-insecurity								0.522***
Techno-uncertainty								
Technostress creators	−0.205**			0.579***				
R2	0.042	0.100	0.165	0.335	0.118	0.375	0.214	0.273
Total F	5.955**	15.085***	26.837***	68.446***	18.231***	81.579***	37.012***	50.944***
Adjusted R2	0.035	0.093	0.159	0.330	0.112	0.370	0.208	0.267

*** $p < 0.01$. ** $p < 0.05$. * $p < 0.1$

borders [19] while working remotely during the COVID-19 lockdown, the chapter adds value to complex understanding of the research constructs and relationships between them.

Third, turning to the impact of individual technostress creators on work-life balance and job burnout, the findings are different supporting the notion that different facets of a multidimensional construct might cause different consequences [18].

As it was predicted, the first technostress creator, namely techno-overload served as a determinant of worse work-life balance and increased job burnout. According to the findings, the respondents felt intense techno-overload (mean 3.38) seeing that working remotely required to work more and faster [49]. As such, worse work-life balance and increased job burnout were outcomes of the situations, where employee efficiency in using information in their work was hampered by the amount of relevant information available to them [9] or working while using ICTs was characterised by multiple tasks and activities that an employee must remember to perform, often in parallel or in rapid succession [17]. The findings supported an earlier study where techno-overload while working remotely led to exhaustion [56].

Turning to the second technostress creator, as predicted, the findings revealed that techno-invasion decreased work-life balance and increased job burnout. The majority of respondents agreed that while working remotely they never felt free of ICTs (mean 3.01). Less time for family issues, the need to sacrifice vacations or weekends or work after official working hours added to the blurred border between work and non-work duties [1]; balance between work and life was damaged and burnout increased. The chapter confirms earlier studies revealing that techno-invasion led to burnout [40] or work-family conflict [45].

Referring to job burnout, the findings demonstrated that techno-complexity and techno-insecurity increased job burnout. Thus, feeling emotionally drained and used up, lack of enthusiasm and doubting about one's contribution at work become proliferate when employees feel insecure about their future employment and constantly have to upgrade their skills.

Contrary to the expectations, techno-complexity and techno-insecurity had no statistically significant relationship with work-life balance, while techno-uncertainty had no statistically significant relationship with work-life balance and job burnout. The argumentation behind this might lie in several reasons. The first reason might be the education level of the sample. Almost all the respondents were employees with university degrees and accordingly they might have felt that their ICTs skills were adequate; they might have not been constantly threatened in their jobs by new technologies or new co-workers with more proficient IT skills; or they might have not needed to invest a huge amount of time to understand the ICT work rules. As such, they probably experienced less difficulties in reconciling the competing demands of work and non-work activities. The second reason might be the generations of respondents [11]. More than a half or 55.2% of respondents belonged to generation Y, while 44.9%—to generation X. Generation X are generally more technologically savvy compared to earlier generations "as they have grown up with a variety of electronic gadgetry for much of their lives" [11, p. 1845]. As regards generation Y, they "were the first wave of the digital generation born into the world of technology.

They are highly qualified in digital knowledge; therefore it is easy for them to quickly acquire the use of new tools and devices in IT" [10, p. 92]. Hence, being "born with ICTs" help to live and work with them.

Fourth, the findings confirm the previous notion found in the literature about the "dark side" of ICTs [47]. Initially, the ICTs were supposed to make our lives easier. However, the reality is controversial and, in many cases, opposite to the intentions [53] as in this case, leading to burnout and worse work-life balance.

Fifth, the chapter contributes to the body of research done during the COVID-19 pandemic and lockdown on employee well-being [45] taking into consideration that similar extreme events could happen in the future.

Practical implications

In addition to the theoretical implications, the research has some managerial implications for practitioners. Given that technostress creators impair employee well-being in terms of job burnout and work-life balance, organisational leaders are invited to design some strategies and take some initiatives, which are concerned with eliminating or reducing technostress creators as such. Already 10 years ago, in 2011, Tarafdar et al. (2011) [67] proposed four kinds of organisational mechanisms that could reduce the conditions for and consequences of technostress: educate employees, assist them if they are having technical problems, involve employees in decisions regarding new technologies, and finally help them learn about innovations (how ICTs drive changes in the routines).

As ICTs are inevitable in contemporary organisations, according to the JD-R model, organisations are encouraged to provide some resources to mitigating the demands while working with ICTs. More specifically, the suggestion for practitioners is to follow a sustainable HRM approach with the focus on long-term employee development, regeneration, and renewal not making them harm [63]. Several practices that might be taken into consideration by practitioners are provided below.

First, employee development should be at the heart of organisation's actions while addressing technostress creators. Sharing knowledge about ICTs, training before new ICT introduction and later upgrading training, and clear manuals serve as the examples of employee education.

Second, it is worth strengthening the cooperation between employees while arranging teamwork for dealing with the ICT use problems. Moreover, good relationships between the IT department and employees are relevant. A mentoring system might be also beneficial.

Third, organisations are encouraged to ensure employee participation. Consultations with employees regarding their opinions about advantages and disadvantages of particular ICTs might strengthen the process of ICT acceptance later.

Fourth, the protocol of communication through ICTs is more than welcome. Such protocol should address highly relevant questions regarding not being "always connected", more specifically stating that employees are not obligated to check email after working hours or on weekends or vacations or that it is forbidden to send professional emails or to call work colleagues, including subordinates, in non-working time.

Summing up, the complex of actions with respect to employee cooperation, employee development, employee involvement in decision-making, and protocol of communication through ICTs might create a synergic effect while dealing with technostress creators.

Limitations

This research has some shortcomings that might be addressed in future studies.

The first concern refers to the sample. The study was conducted in one country. This makes it difficult to generalise the findings. Moreover, the research was conducted during the COVID-19 lockdown and this in turn might also prevent from generalising. In further research, it would be interesting to examine whether the relationship of technostress creators, work-life balance, and job burnout varies across countries and whether this variation depends on specific country-level characteristics.

The second concern is related to the respondents' profile. As the respondents mainly were women with university degrees, further studies are encouraged to selected a more heterogeneous sample.

The third concern deals with the comparative nature of study. For further research, it would be worthwhile to compare the findings from three samples: employees working only remotely, employees working only in the offices, and employees working for organisations offering a mix of remote work and in-office time.

Finally, the research has a limited focus on outcomes of technostress creators. Future studies might include more and different outcomes illustrating employee well-being and performance.

8.9 Conclusions

The aim of the chapter was to reveal the linkage between the construct of technostress creators and its five dimensions (namely techno-overload, techno-invasion, techno-complexity, techno-insecurity, and techno-uncertainty), and work-life balance and job burnout while working remotely during the COVID-19 lockdown. The findings confirmed that technostress creators served as job demands and blurred the border between work and non-work activities while working remotely. More specifically, technostress creators impaired the work-life balance and led to job burnout. Further, the results revealed that not all technostress creators similarly affected the employee well-being. Acknowledging the "dark side" of technologies and seeing that the use of ICTs is inevitable in the contemporary world, organisations are encouraged to deal with techno-overload, techno-invasion, techno-complexity, techno-insecurity, and techno-uncertainty and thus strive for better employee well-being.

References

1. Adisa, T.A., Abdulraheem, I., Isiaka, S.B.: Patriarchal hegemony: investigating the impact of patriarchy on women's work-life balance. Gend. Manag. Int. J. (2019). https://doi.org/10.1108/PR-08-2016-0222
2. Ahmad, U.N.U., Amin, S.M., Ismail, W.K.W.: The relationship between technostress creators and organisational commitment among academic librarians. Procedia Soc. Behav. Sci. **40**, 182–186 (2012). https://doi.org/10.1016/j.sbspro.2012.03.179
3. Ahuja, M.K., Chudoba, K.M., Kacmar, C.J., McKnight, D.H., George, J.F.: IT road warriors: balancing work-family conflict, job autonomy, and work overload to mitigate turnover intentions. Mis Q. 1–17 (2007)
4. Ayyagari, R., Grover, V., Purvis, R.: Technostress: technological antecedents and implications. MIS Q. 831–858 (2011)
5. Bakker, A., Demerouti, E., Schaufeli, W.: Dual processes at work in a call centre: an application of the job demands–resources model. Eur. J. Work Organ. Psy. **12**(4), 393–417 (2003). https://doi.org/10.1080/13594320344000165
6. Bakker, A.B., Demerouti, E.: The job demands-resources model: state of the art. J. Manag. Psychol. (2007). https://doi.org/10.1108/02683940710733115
7. Bakker, A.B., Demerouti, E.: Job demands–resources theory: taking stock and looking forward. J. Occup. Health Psychol. **22**(3), 273 (2017). https://doi.org/10.1037/ocp0000056
8. Barbuto, A., Gilliland, A., Peebles, R., Rossi, N., Shrout, T.: Telecommuting: Smarter Workplaces (2020). http://hdl.handle.net/1811/91648. Accessed 12 Sept 2021
9. Bawden, D., Robinson, L.: The dark side of information: overload, anxiety and other paradoxes and pathologies. J. Inf. Sci. **35**(2), 180–191 (2009). https://doi.org/10.1177/0165551508095781
10. Bencsik, A., Horváth-Csikós, G., Juhász, T.: Y and Z generations at workplaces. J. Compet. **8**(3) (2016). https://doi.org/10.7441/joc.2016.03.06
11. Benson, J., Brown, M.: Generations at work: are there differences and do they matter? Int. J. Human Resour. Manag. **22**(9), 1843–1865 (2011). https://doi.org/10.1080/09585192.2011.573966
12. Borgmann, A.: Technology as a cultural force: for Alena and Griffin. Can. J. Sociol. **31**(3), 351–360 (2006)
13. Brillhart, P.E.: Technostress in the workplace: managing stress in the electronic workplace. J. Am. Acad. Bus. **5**(1/2), 302–307 (2004)
14. Brislin, R.W.: Back-translation for cross-cultural research. J. Cross Cult. Psychol. **1**, 185–216 (1970). https://doi.org/10.1177/135910457000100301
15. Brod, C.: Technostress: The Human Cost of the Computer Revolution. Addison-Wesley, Reading, MA (1984)
16. Brough, P., Timms, C., Bauld, R.: Measuring work-life balance: validation of a new measure across five Anglo and Asian samples. In: Proceedings of the 8th Australian Psychological Society Industrial & Organizational Conference (2009). https://doi.org/10.1080/09585192.2014.899262
17. Cao, H., Lee, C.J., Iqbal, S., Czerwinski, M., Wong, P.N., Rintel, S., Yang, L.: Large scale analysis of multitasking behavior during remote meetings. In: Proceedings of the 2021 CHI Conference on Human Factors in Computing Systems, pp. 1–13 (2021). https://doi.org/10.1145/3411764.3445243
18. Chandra, S., Shirish, A., Srivastava, S.C.: Does technostress inhibit employee innovation? Examining the linear and curvilinear influence of technostress creators. Commun. Assoc. Inf. Syst. **44**(1), 19 (2019)
19. Clark, S.C.: Work/family border theory: a new theory of work/family balance. Human Relat. **53**(6), 747–770 (2000)
20. De Witte, H.: Job insecurity: review of the international literature on definitions, prevalence, antecedents and consequences. SA J. Ind. Psychol. **31**, 1–6 (2005). https://doi.org/10.4102/sajip.v31i4.200

21. Demerouti, E., Bakker, A.B., Nachreiner, F., Schaufeli, W.B.: The job demands-resources model of burnout. J. Appl. Psychol. **86**(3), 499 (2001)
22. Derks, D., Bakker, A.B.: Smartphone use, work–home interference, and burnout: a diary study on the role of recovery. Appl. Psychol. **63**(3), 411–440 (2014)
23. Derks, D., van Duin, D., Tims, M., Bakker, A.B.: Smartphone use and work–home interference: the moderating role of social norms and employee work engagement. J. Occup. Organ. Psychol. **88**(1), 155–177 (2015). https://doi.org/10.1111/joop.12083
24. Dragano, N., Lunau, T.: Technostress at work and mental health: concepts and research results. Curr. Opin. Psychiatry **33**(4), 407–413 (2020)
25. Etikan, I., Musa, S.A., Alkassim, R.S.: Comparison of convenience sampling and purposive sampling. Am. J. Theor. Appl. Stat. **5**(1), 1–4 (2016)
26. Eurofound: International Labour Organization. Working Anytime, Anywhere: The Effects on the World of Work. Publications Office of the European Union, Luxembourg, ILO, Geneva, Switzerland (2017)
27. Felstead, A., Henseke, G.: Assessing the growth of remote working and its consequences for effort, well-being and work-life balance. N. Technol. Work. Employ. **32**(3), 195–212 (2017)
28. Ghislieri, C., Molino, M., Cortese, C.G.: Work and organizational psychology looks at the fourth industrial revolution: how to support workers and organizations? Front. Psychol. **9**, 2365 (2018). https://doi.org/10.3389/fpsyg.2018.02365
29. Greenhaus, J.H., Collins, K.M., Shaw, J.D.: The relation between work–family balance and quality of life. J. Vocat. Behav. **63**(3), 510–531 (2003)
30. Guest, D.E.: Human resource management and employee well-being: towards a new analytic framework. Hum. Resour. Manag. J. **27**(1), 22–38 (2017). https://doi.org/10.1111/1748-8583.12139
31. Hemp, P.: Death by information overload. Harv. Bus. Rev. **87**(9), 82–89 (2009)
32. Hjálmsdóttir, A., Bjarnadóttir, V.S.: "I have turned into a foreman here at home": families and work–life balance in times of COVID-19 in a gender equality paradise. Gend. Work. Organ. **28**(1), 268–283 (2021). https://doi.org/10.1111/gwao.12552
33. Internet in Lithuania is fastest in the world (2021). https://urm.lt/niujorkas/en/news/internet-in-lithuania-is-fastest-in-the-world_1. Accessed 12 Sept 2021
34. Jawahar, I.M., Stone, T.H., Kisamore, J.L.: Role conflict and burnout: the direct and moderating effects of political skill and perceived organizational support on burnout dimensions. Int. J. Stress. Manag. **14**(2), 142 (2007)
35. Kelliher, C., Richardson, J., Boiarintseva, G.: All of work? All of life? Reconceptualising work-life balance for the 21st century. Hum. Resour. Manag. J. **29**(2), 97–112 (2019). https://doi.org/10.1111/1748-8583.12215
36. Khedhaouria, A., Cucchi, A.: Technostress creators, personality traits, and job burnout: a fuzzy-set configurational analysis. J. Bus. Res. **101**, 349–361 (2019)
37. Korunka, C., Vartiainen, M.: Digital technologies at work are great, aren't they? The development of information and communication technologies (ICT) and their relevance in the world of work. In: An Introduction to Work and Organizational Psychology, pp. 102–120 (2017)
38. Lazarus, R.S.: Psychological Stress and the Coping Process. McGraw-Hill, New York (1966)
39. Leung, L., Zhang, R.: Mapping ICT use at home and telecommuting practices: a perspective from work/family border theory. Telemat. Inform. **34**(1), 385–396 (2017)
40. Mahapatra, M., Pati, S.P.: Technostress creators and burnout: a job demands-resources perspective. In: Proceedings of the 2018 ACM SIGMIS Conference on Computers and People Research, pp. 70–77 (2018)
41. Marsh, K., Musson, G.: Men at work and at home: managing emotion in telework. Gend. Work. Organ. **15**(1), 31–48 (2008). https://doi.org/10.1111/j.1468-0432.2007.00353.x
42. Maslach, C., Schaufeli, W.B., Leiter, M.P.: Job burnout. Annu. Rev. Psychol. **52**(1), 397–422 (2001)
43. McGrath, J.: Stress and behavior in organizations. In: Dunnette, M. (ed.) Handbook of Industrial and Organizational Psychology, pp. 1351–1395. Rand-McNally, Chicago (1976)

44. Molino, M., Cortese, C.G., Ghislieri, C.: Unsustainable working conditions: the association of destructive leadership, use of technology, and workload with workaholism and exhaustion. Sustainability **11**(2), 446 (2019). https://doi.org/10.3390/su11020446
45. Molino, M., Ingusci, E., Signore, F., Manuti, A., Giancaspro, M.L., Russo, V., Cortese, C.G.: Wellbeing costs of technology use during Covid-19 remote working: an investigation using the Italian translation of the technostress creators scale. Sustainability **12**(15), 5911 (2020). https://doi.org/10.3390/su12155911
46. Nunnally, J.C.: Psychometric Theory, 2d edn. McGraw-Hill (1978)
47. O'Driscoll, M.P., Biron, C., Cooper, C.L.: Chapter 4: Work-related technological change and psychological well-being. In: Amichai-Hamburger, Y. (ed.) Technology and Psychological Well-Being, pp. 106–139. Cambridge University Press, New York (2009). https://doi.org/10.1108/S1479-3555(2010)0000008010
48. Palumbo, R.: Let me go to the office! An investigation into the side effects of working from home on work-life balance. Int. J. Public Sect. Manag. (2020). https://doi.org/10.1108/IJPSM-06-2020-0150
49. Ragu-Nathan, T.S., Tarafdar, M., Ragu-Nathan, B.S., Tu, Q.: The consequences of technostress for end users in organizations: conceptual development and empirical validation. Inf. Syst. Res. **19**(4), 417–433 (2008). https://doi.org/10.1287/isre.1070.0165
50. Reiter, N.: Work life balance: what DO you mean? The ethical ideology and underpinning appropriate application. J. Appl. Behav. Sci. **43**(2), 273–294 (2007). https://doi.org/10.1177/0021886306295639
51. Roztocki, N., Soja, P., Weistroffer, H.R.: The role of information and communication technologies in socioeconomic development: towards a multi-dimensional framework (2019). https://doi.org/10.1080/02681102.2019.1596654
52. Saim, M.A.S.M., Rashid, W.E.W., Ma'on, S.N.: Technostress creator and work life balance: a systematic literature review. Rom. J. Inf. Technol. Autom. Control **31**(1), 77–88 (2021). https://doi.org/10.33436/v31i1y202106
53. Salanova, M., Llorens, S., Cifre, E.: The dark side of technologies: technostress among users of information and communication technologies. Int. J. Psychol. **48**(3), 422–436 (2013)
54. Salanova, M., Llorens, S., Ventura, M.: Technostress: the dark side of technologies. In: The Impact of ICT on Quality of Working Life, pp. 87–103. Springer, Dordrecht (2014). https://doi.org/10.1007/978-94-017-8854-0_6
55. Salmela-Aro, K., Rantanen, J., Hyvönen, K., Tilleman, K., Feldt, T.: Bergen Burnout Inventory: reliability and validity among Finnish and Estonian managers. Int. Arch. Occup. Environ. Health **84**(6), 635–645 (2011)
56. Sardeshmukh, S.R., Sharma, D., Golden, T.D.: Impact of telework on exhaustion and job engagement: a job demands and job resources model. N. Technol. Work. Employ. **27**(3), 193–207 (2012)
57. Schaufeli, W.B.: Applying the job demands-resources model. Organ. Dyn. **2**(46), 120–132 (2017). https://doi.org/10.1016/j.orgdyn.2017.04.008
58. Schwab, K.: The Fourth Industrial Revolution. Currency (2017)
59. Shoss, M.K.: Job insecurity: an integrative review and agenda for future research. J. Manag. **43**(6), 1911–1939 (2017). https://doi.org/10.1177/0149206317691574
60. Shu, Q., Tu, Q., Wang, K.: The impact of computer self-efficacy and technology dependence on computer-related technostress: a social cognitive theory perspective. Int. J. Hum.-Comput. Interact. **27**(10), 923–939 (2011). https://doi.org/10.1080/10447318.2011.555313
61. Spagnoli, P., Molino, M., Molinaro, D., Giancaspro, M.L., Manuti, A., Ghislieri, C.: Workaholism and technostress during the COVID-19 emergency: the crucial role of the leaders on remote working. Front. Psychol. **11**, 3714 (2020). https://doi.org/10.3389/fpsyg.2020.620310
62. Srivastava, S.C., Chandra, S., Shirish, A.: Technostress creators and job outcomes: theorising the moderating influence of personality traits. Inf. Syst. J. **25**(4), 355–401 (2015). https://doi.org/10.1111/isj.12067
63. Stankevičiūtė, Ž, Savanevičienė, A.: Designing sustainable HRM: the core characteristics of emerging field. Sustainability **10**(12), 4798 (2018). https://doi.org/10.3390/su10124798

64. Tarafdar, M., Pullins, E.B., Ragu-Nathan, T.S.: Technostress: negative effect on performance and possible mitigations. Inf. Syst. J. **25**(2), 103–132 (2015). https://doi.org/10.1111/isj.12042
65. Tarafdar, M., Tu, Q., Ragu-Nathan, T.S.: Impact of technostress on end-user satisfaction and performance. J. Manag. Inf. Syst. **27**(3), 303–334 (2010)
66. Tarafdar, M., Tu, Q., Ragu-Nathan, B.S., Ragu-Nathan, T.S.: The impact of technostress on role stress and productivity. J. Manag. Inf. Syst. **24**(1), 301–328 (2007). https://doi.org/10.2753/MIS0742-1222240109
67. Tarafdar, M., Tu, Q., Ragu-Nathan, T.S., Ragu-Nathan, B.S.: Crossing to the dark side: examining creators, outcomes, and inhibitors of technostress. Commun. ACM **54**(9), 113–120 (2011). https://doi.org/10.1145/1995376.1995403
68. Thulin, E., Vilhelmson, B., Johansson, M.: New telework, time pressure, and time use control in everyday life. Sustainability **11**(11), 3067 (2019). https://doi.org/10.3390/su11113067
69. La Torre, G., Esposito, A., Sciarra, I., Chiappetta, M.: Definition, symptoms and risk of technostress: a systematic review. Int. Arch. Occup. Environ. Health **92**(1), 13–35 (2019)
70. Vilhelmson, B., Thulin, E.: Who and where are the flexible workers? Exploring the current diffusion of telework in Sweden. N. Technol. Work. Employ. **31**(1), 77–96 (2016)
71. Weil, M.M., Rosen, L.D.: TechnoStress: Coping with Technology @Work @Home @Play. Wiley, New York (1997)
72. Welz, C., Wolf, F.: Telework in the European Union. European Foundation for the Improvement of Living and Working Conditions (2010). www.eurofound.europa.eu
73. Wood, J., Oh, J., Park, J., Kim, W.: The relationship between work engagement and work–life balance in organizations: a review of the empirical research. Hum. Resour. Dev. Rev. **19**(3), 240–262 (2020)
74. Yip, B., Rowlinson, S., Siu, O.L.: Coping strategies as moderators in the relationship between role overload and burnout. Constr. Manag. Econ. **26**(8), 871–882 (2008)
75. Zhao, G., Wang, Q., Wu, L., Dong, Y.: Exploring the structural relationship between university support, students' technostress, and burnout in technology-enhanced learning. Asia Pac. Educ. Res. 1–11 (2021). https://doi.org/10.1007/s40299-021-00588-4

Chapter 9
Evaluation Protocols for the Optimization of Water Treatment Plants

Maria Cristina Collivignarelli, Alessandro Abbà, Marco Carnevale Miino, Francesca Maria Caccamo, Silvia Calatroni, and Giorgio Bertanza

Abstract In recent years, water treatment is becoming more important and in the future the number of water treatment plants (WTPs) is destined to increase. However, plants often operate less than optimally from a resource management point of view. This lack of efficiency in WTPs can lead to a drop in pollutants removal yields, an increase in operating costs as well as a waste of resources, in opposition to the concept of circular economy. In this work, a methodological approach useful to identify the critical treatment phases and evaluate the effectiveness of the diverse upgrade interventions has been proposed. Performance indicators useful for the characterization of the WTPs, after a preliminary monitoring phase, are presented. In addition, examples of functionality tests applied to real WTPs are shown.

M. C. Collivignarelli (✉) · M. Carnevale Miino · F. M. Caccamo · S. Calatroni
Department of Civil and Architectural Engineering, University of Pavia, Via Ferrata 3, 27100 Pavia, Italy
e-mail: mcristina.collivignarelli@unipv.it

M. Carnevale Miino
e-mail: marco.carnevalemiino01@universitadipavia.it

F. M. Caccamo
e-mail: francescamaria.caccamo01@universitadipavia.it

S. Calatroni
e-mail: silvia.calatroni01@universitadipavia.it

M. C. Collivignarelli
Interdepartmental Centre for Water Research, University of Pavia, Via Ferrata 3, 27100 Pavia, Italy

A. Abbà · G. Bertanza
Department of Civil, Environmental, Architectural Engineering and Mathematics, University of Brescia, Via Branze 43, 25123 Brescia, Italy
e-mail: alessandro.abba@unibs.it

G. Bertanza
e-mail: giorgio.bertanza@unibs.it

L. Ivascu et al. (eds.), *Intelligent Techniques for Efficient Use of Valuable Resources*, Intelligent Systems Reference Library 227,
https://doi.org/10.1007/978-3-031-09928-1_9

Keywords Water treatment · Treatment plants efficiency · Treatment plants monitoring · Hydrodynamic anomalies · Performance indicators

9.1 Introduction

In recent years, following the increase of population linked to a sewerage system and the more stringent discharge limits, the number of wastewater treatment plants (WWTPs) is constantly growing [1]. Even in the field of drinking water, the number of drinking water treatment plants (DWTPs) is increasing [2]. The presence of water treatment plants (WTPs) inevitably leads to the consumption of energy and the production of waste and residues (e.g., biological, and chemical sludge) which, if not recovered, must be properly disposed of [3–7].

The European Union, through the recent Green Deal [8], is paying attention to the management of the environmental sector in a more sustainable way. This aspect had already been addressed during the ambitious circular economy action plan [9]. In the transition to more sustainable development, WTPs can play a key role. For instance, in recent years, several researchers theorized a change in the perspective with which to intend the WWTPs no longer as a simple treatment plant to remove pollutants but as a real site for the recovery of resources contained in the wastewater (WW) [10–13]. This change of perspective regards the same substances that according to the old approach were pollutants that need to be eliminated (e.g., organic carbon, nitrogen, phosphorus), as precious substances to be recovered/reused.

Even for drinking water treatment plants, the processing margins are more limited, although present. In this case the research is focusing on increasing efficiency to limit the production of waste by stimulating the recovery of residues, where possible [2, 14].

To ensure that DWTPs and WWTPs are congruent with the concept of a circular economy, the first required step is improving the efficiency of the operation of the plants themselves. In this respect, the monitoring of DWTPs and WWTPs represent an essential step.

In this work, the aspects about the plant monitoring plans, the functional checks, and the performance indicators (PIs) are reported and discussed to highlight the importance of following a precise methodological protocol to optimize and improve DWTPs and WWTPs operation.

9.2 The Importance of Monitoring Plans and Consequent Calculation of Performance Indicators

The monitoring of WTPs represents the starting point thanks to which data can be found for future assessments.

The monitoring plan must consider various aspects such as: (i) the size of the WTPs, (ii) the type of processes that compose them, (iii) the characteristics of the water treated, and (iv) the eventual presence of previous problems in the system. Monitoring activities can be classified as routinary or intensive. The former ones are carried out with a certain periodicity regardless of the operating conditions of the system. The later ones can be undertaken if performance deficiencies in one or more sectors are highlighted, or there is a need to focus on a particular phenomenon.

In any case, the primary aspect of monitoring is to collect the data necessary for subsequent reprocessing aimed at defining the state of "health" of the WTP. The monitoring plan plays a key role in the management of WTPs because it must balance the need to find data with the economic and operational resources available to the water utility.

The data collected during the monitoring phase represent the starting point for the evaluation of the performance indicator, which is useful for understanding not only if the WTP is operating effectively in terms of removing pollutants but also if it is doing it with an acceptable use of materials and energy.

For example, Collivignarelli et al. [15] applied 74 different PIs to evaluate the degree of operation of a conventional WWTP and the sewer system in an integrated manner, focusing not only on operational and functional aspects (e.g., removal of pollutants), but also indicators which took into consideration economic and financial aspects, environmental and acoustic impact, the use of dedicated staff, to present an overall "state of the art" of the degree of functioning of a WTP.

Sabia et al. [16] studied 10 WWTPs of different sizes and proposed a methodology to evaluate the energy performance of the diverse WTPs to provide the water utilities with a useful tool for the rapid identification of the plants where to focus the interventions in a priority manner.

Silva and Rosa [17] analysed the performance of 23 different WTPs with diverse PIs to determine the effectiveness, reliability, overall sludge production and energy efficiency of each plant. Also, their study confirmed the feasibility to use the PIs as a useful evaluation and planning tool in the management of WTPs.

9.3 Functionality Checks

Functionality checks help to determine the efficiency of an entire system, e.g. conventional activate sludge (CAS) system or a single reactor. The checks can be classified into conventional or innovative ones [18]. The first are those that have been known for several years (for instance, check of reactors' hydrodynamic, study of sludge settleability, determination of oxygen transfer capacity) while in the second category those applied only a few years in this context (e.g., respirometric tests) [19]. Although known from several years, these checks are little applied in the ordinary WTPs management, so far [18, 19].

About WW, the Italian legislation poses both a threshold limit to the pollutants concentration discharged in the receiving water body that a minimum values to the percentages of removal of the same [20], providing implicitly that the WWTPs should reduce effectively the concentration of organic substance and nutrients present in WW. However, to date, there is not a specific national or European legislation concerning the minimum requirements of efficiency of WWTPs. With reference to this aspect, the Lombardy Region (Italy) has recently introduced a legislation [21] that makes the periodic WWTPs functionality checks mandatory, in order to monitor the efficiency of the plants and to intervene promptly where necessary. Instead, concerning the DWTPs, the legislation does not currently impose to carry out functionality tests.

The hydrodynamic tests are among the more applied features of functionality and among this residence time distribution (RTD) and computational fluid dynamics (CFD) represent the most common analyses [19, 22, 23]. The RTD analysis includes a first experimental part in which a tracer is released (e.g., lithium chloride) and its concentration is monitored over time, and a second phase in which the system response curve is compared with model curves to determine the type of hydrodynamic anomaly (by-pass of flow rate or dead volume) and quantify its entity [19, 24, 25].

Collivignarelli et al. [26] applied three RTD analysis on a CAS system, in which nitrification and denitrification were also present, to assess the effectiveness of corrective interventions implemented by the manager of WWTP thorough several years and the effects of the accumulation of inert materials on the hydrodynamics of the reactors. The results showed that both reactors presented significant hydrodynamic anomalies both in terms of flow bypass and dead volumes. These problems were than solved thanks to the corrective interventions implemented by the manager of WWTP, specifically the change of the position of submerged mixers, the introduction of two separate flows for the recirculation of mixed liquor, and the sinking of sludge recirculation pipe below the surface of water [26].

Instead, Aral and Demirel [27] applied the RTD analysis to test a particular design for chlorine or ozone disinfection tanks in DWTPs to minimize hydrodynamic anomalies and therefore also the management costs of the plant (e.g., costs of reagents).

The CFD analysis allows to overcome the main disadvantages of the RTD analysis such as the impossibility of determining the precise localization of the anomalies but requires important calculation charges in the case of a too dense calculating mesh [19, 24]. In recent years, several authors have therefore proposed the use of RTD and CFD in a combined analysis.

Manenti et al. [24] employed this approach on a CAS pilot plant observing that the quantification of the hydrodynamic anomalies of the two analyses was comparable and perfectly integrated. In another recent work, Collivignarelli et al. [19] applied integrated RTD-CFD analysis to a real scale WWTP confirming the full compatibility of the two analyses. The RTD system was able to perfectly quantify the dead volumes present in the biological reactors, while the CFD analysis allowed to determine the position of the anomalies using a reduced precision mesh to limit the calculation charges.

9.4 Upgrade of the Plant: How Performance Indicators Can Help in the Decision?

Based on the results of the functionality tests, decisions regarding the upgrade of the WTPs can be made. The upgrades may involve already existing sectors (functional upgrade, e.g., removal of debris accumulated in the reactors, implementation of new mechanical mixing systems, …), or they may include the construction of new reactors and/or the integration of new processes to existing ones (structural upgrades).

To identifying the most appropriate choice in the decision-making operations prior to each upgrade, the calculation of the PIs can be a valid tool in defining the choice.

The knowledge of WTP conditions helps to evaluate in which process to act on by establishing an order of priority of intervention, especially considering limited economic resources. By estimating the effects of the upgrades, thanks also to laboratory and pilot scale tests, it is possible to predict the impact they will have on the overall performance of the WTP, helping the water utilities to classify the intervention in order of priority [15].

9.5 Discussion

Proper management of WTPs represents a key aspect to obtain an effective water treatment action, limiting the waste of material, energy, and economic resources.

In Fig. 9.1, a methodological approach applicable to WTPs is reported. Monitoring, through the drafting of an adequate protocol, plays a key role for all subsequent steps of the methodological approach. Indeed, the calculation of the PIs of the diverse WTPs is possible only thanks to the data obtained through a capillary and effective monitoring phase. If the results are satisfactory, this represents an indicator of the effective and efficient functioning of the WTP. However, the indices must be periodically recalculated to determine if there are any discrepancies with respect to the values previously obtained and therefore to identify any critical phases of the WTP.

If the results are not satisfactory, it is possible to act with functionality checks to determine, quantify, locate, and solve the problem. On the contrary, if the results are satisfactory, it indicates that the functionality of the processes is acceptable and therefore the low effectiveness of the WTP is not given by poor process efficiency. In this case, structural upgrades (construction of new reactors, introduction of new processes) are recommended. If the results are not satisfactory, management upgrades are necessary to correct the functionality of the existing processes and bring it back to an optimal level. If this is not enough to guarantee the complete effectiveness of the WTP, it will be necessary to adopt structural upgrades.

The size of the WTPs, represents one of the limits of this approach. Potentially, in the presence of a sufficient amount of monitoring data, the approach would be applicable to all types of WTPs. However, in practice, the monitoring associated

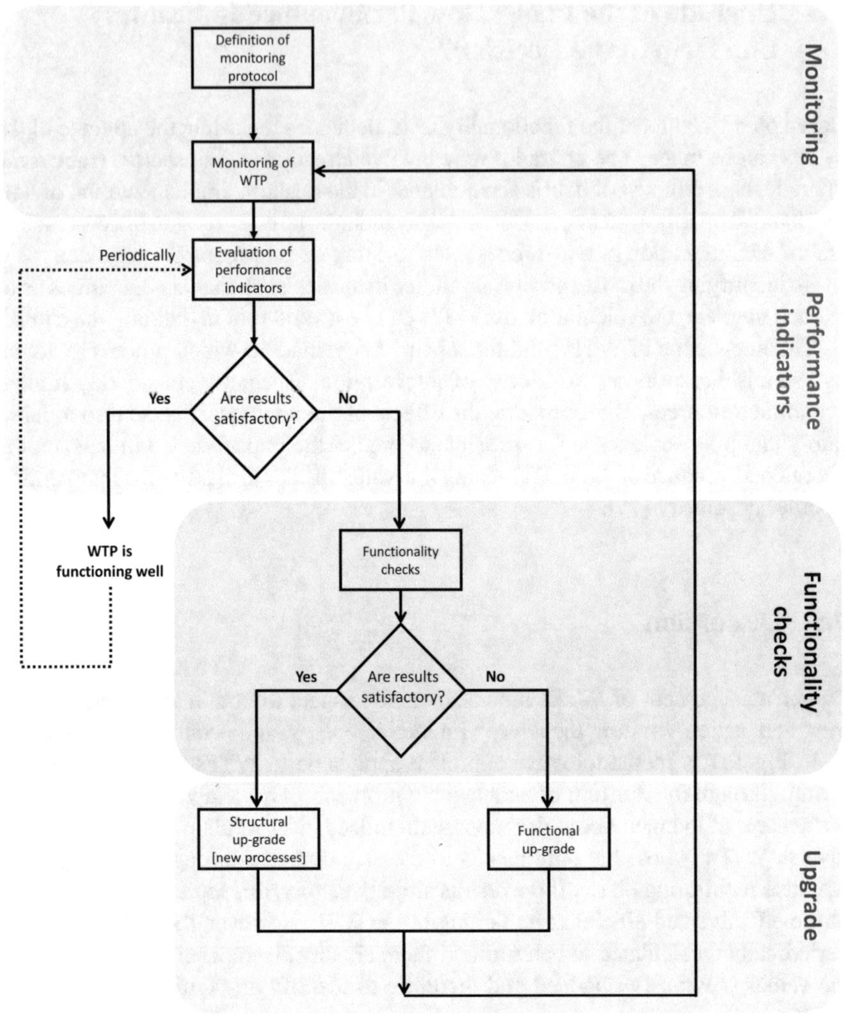

Fig. 9.1 Proposal of methodological approach for optimal WTPs management

with low-population equivalent WTPs is often not very detailed and sparse. This leads to the impossibility of calculating part of the PIs, with a consequent only partial evaluation of the degree of effectiveness and efficiency of the WTPs. However, in these cases it is possible to intervene directly by carrying out checks to determine the functionality of the systems and implement functional or structural upgrade solutions, if necessary.

9.6 Conclusions

The lack of efficiency in WTPs can drop pollutants removal yields and increase in operating costs. In this work, a methodological approach has been proposed. Each manager of the water utilities can apply the protocol to verify the performance of the service over its lifecycle, highlighting the improvements, critical issues, and strengths. The protocol is composed by diverse steps: (i) a preliminary monitoring phases, (ii) the calculation of PIs, (iii) the carrying out of functionality checks, and (iv) the upgrade of WTPs with functional or structural interventions. This approach acts as a simple and useful tool for the correct management of WTPs to limit consumption of reagents and energy while guaranteeing high performance.

References

1. Lu, J.-Y., Wang, X.-M., Liu, H.-Q., Yu, H.-Q., Li, W.-W.: Optimizing operation of municipal wastewater treatment plants in China: the remaining barriers and future implications. Environ. Int. **129**, 273–278 (2019). https://doi.org/10.1016/j.envint.2019.05.057
2. Sorlini, S., Collivignarelli, M.C., Castagnola, F., Crotti, B.M., Raboni, M.: Methodological approach for the optimization of drinking water treatment plants' operation: a case study. Water Sci. Technol. **71**, 597–604 (2015). https://doi.org/10.2166/wst.2014.503
3. Collivignarelli, M.C., Canato, M., Abbà, A., Carnevale Miino, M.: Biosolids: what are the different types of reuse? J. Clean. Prod. **238**, 117844 (2019). https://doi.org/10.1016/j.jclepro.2019.117844
4. Gherghel, A., Teodosiu, C., De Gisi, S.: A review on wastewater sludge valorisation and its challenges in the context of circular economy. J. Clean. Prod. **228**, 244–263 (2019). https://doi.org/10.1016/j.jclepro.2019.04.240
5. Collivignarelli, M.C., Cillari, G., Ricciardi, P., Carnevale Miino, M., Torretta, V., Rada, E.C., Abbà, A.: The production of sustainable concrete with the use of alternative aggregates: a review. Sustainability **12**, 7903 (2020). https://doi.org/10.3390/su12197903
6. Collivignarelli, M.C., Abbà, A., Frattarola, A., Carnevale Miino, M., Padovani, S., Katsoyiannis, J., Torretta, V.: Legislation for the reuse of biosolids on agricultural land in Europe: overview. Sustainability **11**, 6015 (2019). https://doi.org/10.3390/su11216015
7. Collivignarelli, M.C., Abbà, A., Carnevale Miino, M., Cillari, G., Ricciardi, P.: A review on alternative binders, admixtures and water for the production of sustainable concrete. J. Clean. Prod. **295**, 126408 (2021). https://doi.org/10.1016/j.jclepro.2021.126408
8. EC A European Green Deal—striving to be the first climate-neutral continent. https://ec.europa.eu/info/strategy/priorities-2019-2024/european-green-deal_en
9. European Union Circular Economy Action Plan
10. Regmi, P., Stewart, H., Amerlinck, Y., Arnell, M., García, P.J., Johnson, B., Maere, T., Miletić, I., Miller, M., Rieger, L., Samstag, R., Santoro, D., Schraa, O., Snowling, S., Takács, I., Torfs, E., van Loosdrecht, M.C.M., Vanrolleghem, P.A., Villez, K., Volcke, E.I.P., Weijers, S., Grau, P., Jimenez, J., Rosso, D.: The future of WRRF modelling—outlook and challenges. Water Sci. Technol. **79**, 3–14 (2019). https://doi.org/10.2166/wst.2018.498
11. Coats, E.R., Wilson, P.I.: Toward nucleating the concept of the water resource recovery facility (WRRF): perspective from the principal actors. Environ. Sci. Technol. **51**, 4158–4164 (2017). https://doi.org/10.1021/acs.est.7b00363
12. Seco, A., Aparicio, S., González-Camejo, J., Jiménez-Benítez, A., Mateo, O., Mora, J.F., Noriega-Hevia, G., Sanchis-Perucho, P., Serna-García, R., Zamorano-López, N., Giménez,

J.B., Ruiz-Martínez, A., Aguado, D., Barat, R., Borrás, L., Bouzas, A., Martí, N., Pachés, M., Ribes, J., Robles, A., Ruano, M.V., Serralta, J., Ferrer, J.: Resource recovery from sulphate-rich sewage through an innovative anaerobic-based water resource recovery facility (WRRF). Water Sci. Technol. **78**, 1925–1936 (2018). https://doi.org/10.2166/wst.2018.492

13. Forouzanmehr, F., Le, Q.H., Solon, K., Maisonnave, V., Daniel, O., Buffiere, P., Gillot, S., Volcke, E.I.P.: Plant-wide investigation of sulfur flows in a water resource recovery facility (WRRF). Sci. Total Environ. **801**, 149530 (2021). https://doi.org/10.1016/j.scitotenv.2021.149530
14. Farhaoui, M., Derraz, M.: Review on optimization of drinking water treatment process. J. Water Resour. Prot. **08**, 777–786 (2016). https://doi.org/10.4236/jwarp.2016.88063
15. Collivignarelli, M.C., Todeschini, S., Abbà, A., Ricciardi, P., Carnevale Miino, M., Torretta, V., Rada, E.C., Conti, F., Cillari, G., Calatroni, S., Lumia, G., Bertanza, G.: The performance evaluation of wastewater service: a protocol based on performance indicators applied to sewer systems and wastewater treatment plants. Environ. Technol. 1–18 (2021). https://doi.org/10.1080/09593330.2021.1922509
16. Sabia, G., Petta, L., Avolio, F., Caporossi, E.: Energy saving in wastewater treatment plants: a methodology based on common key performance indicators for the evaluation of plant energy performance, classification and benchmarking. Energy Convers. Manag. **220**, 113067 (2020). https://doi.org/10.1016/j.enconman.2020.113067
17. Silva, C., Rosa, M.J.: Performance assessment of 23 wastewater treatment plants—a case study. Urban Water J. **17**, 78–85 (2020). https://doi.org/10.1080/1573062X.2020.1734634
18. Collivignarelli, M.C., Abbà, A., Bertanza, G., Damiani, S., Raboni, M.: Resilience of a combined chemical-physical and biological wastewater treatment facility. J. Environ. Eng. **145**, 05019002 (2019). https://doi.org/10.1061/(ASCE)EE.1943-7870.0001543
19. Collivignarelli, M.C., Carnevale Miino, M., Manenti, S., Todeschini, S., Sperone, E., Cavallo, G., Abbà, A.: Identification and localization of hydrodynamic anomalies in a real wastewater treatment plant by an integrated approach: RTD-CFD analysis. Environ. Process. **7**, 563–578 (2020). https://doi.org/10.1007/s40710-020-00437-4
20. Government of Italy Legislative Decree 3 April 2006, n. 152. Environmental regulations (in Italian)
21. LR Regional Regulation 29 March 2019, n. 6.—Discipline and administrative regimes of discharges of domestic wastewater and urban wastewater, regulation of the controls of discharges and the methods of approval of projects for urban wastewater treatment plants, in implementation of Article 52, paragraphs 1, letters a) ef bis), and 3, as well as article 55, paragraph 20, of the regional law 12 December 2003, n. 26 (Discipline of local services of general economic interest. Rules on waste management, energy, use of the subsoil and water resources) (in Italian); Milan, Italy, 2019; p BURL n. 14 suppl
22. Sánchez, F., Viedma, A., Kaiser, A.S.: Hydraulic characterization of an activated sludge reactor with recycling system by tracer experiment and analytical models. Water Res. **101**, 382–392 (2016). https://doi.org/10.1016/j.watres.2016.05.094
23. Sarkar, M., Sangal, V.K., Bhunia, H.: Hydrodynamics and parametric study of an activated sludge process using residence time distribution technique. Environ. Eng. Res. **25**, 400–408 (2019). https://doi.org/10.4491/eer.2019.114
24. Manenti, S., Todeschini, S., Collivignarelli, M.C., Abbà, A.: Integrated RTD—CFD hydrodynamic analysis for performance assessment of activated sludge reactors. Environ. Process. **5**, 23–42 (2018). https://doi.org/10.1007/s40710-018-0288-5
25. Raboni, M., Gavasci, R., Viotti, P.: Influence of denitrification reactor retention time distribution (RTD) on dissolved oxygen control and nitrogen removal efficiency. Water Sci. Technol. **72**, 45–51 (2015). https://doi.org/10.2166/wst.2015.188
26. Collivignarelli, M.C., Bertanza, G., Abbà, A., Damiani, S.: Troubleshooting in a full-scale wastewater treatment plant: what can be learnt from tracer tests. Int. J. Environ. Sci. Technol. **16**, 3455–3466 (2019). https://doi.org/10.1007/s13762-018-2032-0
27. Aral, M.M., Demirel, E.: Novel slot-baffle design to improve mixing efficiency and reduce cost of disinfection in drinking water treatment. J. Environ. Eng. **143**, 06017006 (2017). https://doi.org/10.1061/(ASCE)EE.1943-7870.0001266

Chapter 10
Learning Model-Free Reference Tracking Control with Affordable Systems

Mircea-Bogdan Radac and Alexandra-Bianca Borlea

Abstract This chapter validates and discusses the application of two intelligent learning control techniques, namely Model-free Value Iteration Reinforcement Learning (MFVIRL) and Virtual State-feedback Reference Tuning (VSFRT), for linear output reference model (ORM) tracking of three inexpensive lab scale systems which are interacted with by the help of modern software and hardware. The lab systems consist of an Electrical Braking System (EBS) emulator which is a very representative resistive-based dissipative device, a lab scale Active Temperature Control System (ACTS) which is another area of interest for home/industrial applications and a generic Voltage Control Electrical System (VCES) device, as representative complex, nonlinear and multidimensional systems. The control techniques are unique in the sense that they pertain to different paradigms such as artificial intelligence and classical control, however they share the same learning goal formulation. Herein, learning is based on a virtual state representation built from input–output (I/O) data measured from the systems by interaction, under exploration settings. The learned linear feedback controllers indirectly linearize the closed-loop system, by adequate selection of the linear ORM whose behavior is to be replicated by the closed-loop.

Keywords Learning systems · Intelligent control · Reinforcement learning · Virtual state-feedback reference tuning · Output reference model tracking · Neural networks · Electrical braking system · Active temperature control system · Voltage control electrical system · Input–output observability

10.1 Introduction

Research in data-driven control is strongly motivated by the idea that the main factor responsible for control quality degradation is the mismatch between the identified

M.-B. Radac (✉) · A.-B. Borlea
Department of Automation and Applied Informatics, Politehnica University of Timisoara, 2, Vasile Parvan Ave., 300223 Timisoara, Romania
e-mail: mircea.radac@upt.ro

L. Ivascu et al. (eds.), *Intelligent Techniques for Efficient Use of Valuable Resources*, Intelligent Systems Reference Library 227,
https://doi.org/10.1007/978-3-031-09928-1_10

system model used for model-based control design and the true system. To overcome this problem, learning control directly from input–output (I/O) data has become a prevalent topic in recent years, with numerous paradigms proposed. In this chapter, two important paradigms are investigated, to solve the same model reference control problem, herein called the output reference model (ORM) tracking problem.

The ORM tracking or matching problem is a well-known concept for which most developments originated with the model reference adaptive control (MRAC) classical control field. When a linear ORM is sought to be matched by the closed-loop control system (CLS) by a proper control design, the most interesting and common situation is when the controlled system is nonlinear. Hence a nonlinear controller is required to ensure that the CLS matches the linear ORM, ensuring indirect feedback linearization. To date, output-feedback and state-feedback are the known concepts to achieve this indirect feedback linearization. The state-feedback control is in many ways superior as it uses supplementary information, however its drawbacks reside with state information availability: sensors are expensive and since they are many, they have a lower fault tolerance. While state estimation is usually model-based and requires prior information about the controlled system. This is where the observability assumption comes into context as it proposes a state representation consisting of sequences of I/O data samples. This fact is known for a long time in linear systems theory and also been exploited in nonlinear systems' control as well [1–5]. This virtual state representation from I/O data is then used for deriving a kind of virtual state-feedback control. The ORM tracking concept has been employed, e.g. in [2, 6, 7], as a low-level building block in more advanced hierarchical control systems, [8–12], confirming its practical necessity.

The reinforcement learning (RL) control also known as approximate (or adaptive) dynamic programming (ADP) control is an ongoing research topic with simultaneous developments in the artificial intelligence and control communities. Much of the RL approach in classical control is due to some earlier works by [13–16] and further formalized in [17–26]. In the control systems area, the two most popular algorithms for solving RL-based control are Value Iteration (VI) and Policy Iteration (PoIt). Many variants of VI and PoIt have been proposed, offline or online, model-based or model-free, batch- or adaptive-wise. Subjectively, the most important issues with the mode-free variants of RL-based algorithms (which are mostly interesting due to their independence with respect to system dynamics knowledge) are: the state encoding, the exploration–exploitation dilemma and the suitable parameterization for the cost (or value) function and for the controller, respectively. The last two aspects mainly define the learning data-efficiency of the RL algorithms. RL is not limited to learning ORM control, but to wider learning contexts. However, in this work it will be used for the first mentioned context, under a model-free VI RL (MFVIRL) implementation variant. And it will be developed for a fully state-observable environment, due to the virtual state representation based on I/O data samples. Such representation has been used before in artificial intelligence applications of RL, in order to encode the state information, but without a proper theoretical foundation [3].

While the model-free Virtual Reference Feedback Tuning (VRFT) is another paradigm which is more well-known to the classical control practitioners [27–35]. It is related to other similar data-driven methods stemming from classical control such as IFT [36], SPSA [37], CbT [38], Unfalsified Control [39], Extremum Seeking [40], ILC [41–44], Model-Free (Adaptive) Control [45, 46], etc. Classical VRFT is based on output feedback to ensure ORM tracking. Differently from RL, VRFT was aimed in the first place to ensure this learning goal, hence it can be considered of narrower scope. The main issues with VRFT are related to somewhat classical control aspects such as reference model selection (bandwidth, time-delay, nonminimum phase character), learning efficiency combined with the exploration issue (called persistency of excitation in the classical control), and finally the controller complexity in terms of parameterization. Within this work, the state-feedback variant of VRFT called Virtual State Feedback Reference Tuning (VSFRT) is of primordial concern in the context of learning ORM tracking.

The two above-mentioned control approaches share the same learning objective and some form of hybridization or inter-dependent usage [35, 47, 48] has emerged. In some works, VRFT (VSFRT) has been used to develop a priori stabilizing controllers for enhancing the exploration efficiency, in order to collect more input-state-output data required by the more data-hungry MFVIRL. While in others, VRFT/VSFRT were used to initialize the controller for the MFVIRL algorithm, therefore expected to shorten the learning convergence in terms of fewer iterations and less time [47–49].

This chapter aims at proving that model-free control techniques such as MFVIRL and VSFRT are applicable to affordable, cheap control systems of different natures, which can be interacted with using modern and widely-available software such as MATLAB, to harness the power of available and popular hardware such as e.g. Arduino, in order to obtain complex, nonlinear, multivariable systems. Therefore, these intelligent techniques discuss elements of efficient use of valuable resources, integrating data with knowledge to aim for the next generation of adaptive and intelligent control systems.

Section 10.2 introduces the ORM tracking problem and proposes solutions with MFVIRL and with VSFRT intelligent learning control techniques. Two algorithms synthesize the application steps of these two techniques, for practicality reasons. Section 10.3 describes the application of MFVIRL and VSFRT to an Electrical Braking System (EBS) emulator which is a very representative resistive-based dissipative device widespread found in cars and trains but also in wind turbine generators ensembles. Section 10.4 validates the two aforementioned techniques on a lab scale Active Temperature Control System (ACTS) which is another area of interest for home/industrial applications. While Sect. 10.5 validates the control learning approaches on a generic Voltage Control Electrical System (VCES) device. The concluding Sect. 10.6 summarizes the contributions of the chapter and proposes several research directions.

Notation: From a notation viewpoint, most terminology should be familiar to the control system practitioners. We emphasize the use of boldface notation ($\boldsymbol{u}$, $\boldsymbol{y}$, $\boldsymbol{F}$, $\boldsymbol{P}$, etc.) for vectors and matrices (and functions of vectors and matrices, respectively), while the terms without bolded notation denote scalars. This is irrespective of the lowercase or uppercase usage. $\mathfrak{R}$ denotes the set of real numbers, $\mathfrak{R}^n$ are n-dimensional vectors and $\mathfrak{R}^{m \times n}$ are real matrices with m lines by n columns. The convention for vectors is column notation in this work. Also, the norm operator $||.||$ measures the Euclidean distance applied to vectors.

10.2 The Linear ORM Tracking Control

10.2.1 The Model Reference Tracking Formulation

A discrete-time I/O equation representing the system is (with k set subscript as the sample index and using implicit column vector notation)

$$\boldsymbol{y}_k = \boldsymbol{f}\left(\boldsymbol{y}_{k-1}, \ldots, \boldsymbol{y}_{k-ny}, \boldsymbol{u}_{k-1}, \ldots, \boldsymbol{u}_{k-nu}\right), \tag{10.1}$$

having $\boldsymbol{u}_k = \left[u_{k,1}, \ldots, u_{k,d_u}\right]^T \in \Omega_U \subset \Re^{d_u}$ as the d_u-dimensional system control input within the known domain Ω_U, $\boldsymbol{y}_k = \left[y_{k,1}, \ldots, y_{k,d_y}\right]^T \in \Omega_Y \subset \Re^{d_y}$ as the d_y-dimensional system output within the known domain Ω_Y, $ny, nu \in \mathbb{Z}_+^*$ are unknown orders and the nonlinear I/O system function $\boldsymbol{f} : \Omega_Y \times \cdots \times \Omega_Y \times \Omega_U \times \cdots \times \Omega_U \rightarrow \Omega_Y$ is continuously differentiable. The nonlinear dynamics captured by (10.1) has two basic data-driven assumptions: it is both controllable and observable, at least within the domains of interest (which means locally). Although untestable from the system's unknown function, these assumptions are based on work experience with the system, from datasheets or from other technical documentation. Most of the system that we interact with are controllable, otherwise they would not represent an interest from this perspective. While observability is a deeper, more subtle requirement. Observability property implies that, under free system response (i.e. independent of the trajectory $\boldsymbol{u}_k$), the output trajectory $\boldsymbol{y}_k$ can be used to determine the unique state trajectory of (10.1). This state trajectory, let us call it "$\boldsymbol{x}_t$", is not explicit however. Then, a state-space model is required. Most of this interpretation is well-known in linear systems theory but has been exploited with nonlinear systems as well, where most of the concepts extrapolate easily. Relying upon the nonlinear observability assumption, the work [1] proposes Lemma 1 to build a state-space model acting as an alias (or a virtual system) for model (10.1)

$$\boldsymbol{x}_{k+1} = \boldsymbol{F}(\boldsymbol{x}_k, \boldsymbol{u}_k), \qquad \boldsymbol{y}_k = \left[\mathbf{1}_{d_y}, 0, \ldots, 0\right]\boldsymbol{x}_k = \boldsymbol{x}_{k,1}, \tag{10.2}$$

with notations being detailed in the following: $\mathbf{1}_{d_y} = [1, \ldots, 1] \in \Re^{d_y}$, the virtual state vector accumulates system's (10.1) I/O data samples as described by $\boldsymbol{x}_k = [\boldsymbol{P}_{Y\overline{k,k-\tau}}^T, \boldsymbol{P}_{U\overline{k-1,k-\tau}}^T]^T \triangleq \left[\boldsymbol{x}_{k,1}^T, \boldsymbol{x}_{k,2}^T, \ldots \boldsymbol{x}_{k,2\tau+1}^T\right]^T \in \Omega_X \subset \Re^{(\tau+1)d_y+\tau d_u}$, i.e. with partial vectors $\boldsymbol{P}_{Y\overline{k,k-\tau}} = [\boldsymbol{y}_k^T, \ldots, \boldsymbol{y}_{k-\tau}^T]^T = [\boldsymbol{x}_{k,1}^T, \ldots, \boldsymbol{x}_{k,\tau+1}^T]^T \subset \Re^{(\tau+1)d_y}$, $\boldsymbol{P}_{U\overline{k-1,k-\tau}} = [\boldsymbol{u}_{k-1}^T, \ldots, \boldsymbol{u}_{k-\tau}^T]^T = [\boldsymbol{x}_{k,\tau+2}^T, \ldots, \boldsymbol{x}_{k,2\tau+1}^T]^T \subset \Re^{\tau d_u}$, and the partially unknown nonlinear system function is $\boldsymbol{F}(.) : \Omega_X \times \Omega_U \rightarrow \Omega_X$ (it is Ref. [1] that argues why $\boldsymbol{F}(.)$ is partially unknown). However, the domain Ω_X of the virtual state is known since Ω_U, Ω_Y are known. A nonlinear observability index $\tau \in \mathbb{Z}_+^*$ whose role is identical to the linear systems observability index, defines here as the smallest strictly positive integer for which the state $\boldsymbol{x}_k$ from (10.2) is expressible in terms of the I/O samples $\boldsymbol{u}_k$, $\boldsymbol{y}_k$ from system (10.1) (see Ref. [2]). Meaning that, although not explicitly stated, a nonlinear state-space representation for (10.1) is feasible per (10.2). With unknown orders ny, nu in (10.1), its state space dimension is also unknown, meaning that so it is τ. To select τ, one searches for the optimal value ensuring the best control based on the state information $\boldsymbol{x}_k$. This search is commonly performed by trial and error.

Observation 1. The same input and output apply to both (10.1) and (10.2), so their I/O control is equivalent. The intent is to use the state $\boldsymbol{x}_k$ from (10.2) for feedback purposes, to develop a control law for calculating $\boldsymbol{u}_k$ in (10.1). With all state components measurable, (10.2) is completely state-observable.

Observation 2. Model (10.1) can have time delays included and still transform to (10.2) by defining extra states. Such time delays are identifiable from historical I/O data recordings.

To define the model reference control tracking problem, we start with the linear ORM (LORM) definition in the state-space form

$$\begin{cases} \boldsymbol{x}_{k+1}^m = \boldsymbol{A}_{ORM}\boldsymbol{x}_k^m + \boldsymbol{B}_{ORM}\boldsymbol{\rho}_k, \\ \boldsymbol{y}_k^m = \boldsymbol{C}_{ORM}\boldsymbol{x}_k^m, \end{cases} \tag{10.3}$$

acknowledging that (10.3) is in fact a state-space realization of the linear pulse transfer matrix (PTM) $\boldsymbol{M}(q)$ (q is the time-step forward operator) which relates $\boldsymbol{y}_k^m = \boldsymbol{M}(q)\boldsymbol{\rho}_k$, over infinitely many realizations $(\boldsymbol{A}_{ORM}, \boldsymbol{B}_{ORM}, \boldsymbol{C}_{ORM})$. Herein, the LORM state defines in vector notation as $\boldsymbol{x}_k^m = [x_{k,1}^m, \ldots, x_{k,n_x}^m] \in \Omega_{X^m} \subset \mathfrak{R}^{n_x}$, the reference input $\boldsymbol{\rho}_k = \left[\rho_{k,1}, \ldots, \rho_{k,d_y}\right]^T \in \Omega_\rho \subset \mathfrak{R}^{d_y}$ is a vector within known domain Ω_ρ and the LORM's output is $\boldsymbol{y}_k^m = \left[y_{k,1}^m, \ldots, y_{k,d_y}^m\right]^T \in \Omega_{Y^m} \subset \mathfrak{R}^{d_y}$ is a vector within known domain. $\boldsymbol{M}(q)$'s choice is seldom uninformed since the model reference control paradigm indicates that $\boldsymbol{M}(q)$ should correlate with system's (10.1) bandwidth, while including its time delay and nonminimum-phase character, if any. Usually, unit-gain is required for the PTM $\boldsymbol{M}(q)$, making $\Omega_\rho = \Omega_{Y^m}$ overlapped. In regard to these qualitative system indices, some previous experience is necessary from the designer's part, to be able to interpret these indices and to be able to properly choose the LORM.

We formally define the LORM tracking (or matching) control problem as the following optimization

$$\begin{aligned} &\boldsymbol{u}_0^*, \boldsymbol{u}_1^*, \cdots = arg \min_{\boldsymbol{u}_0, \boldsymbol{u}_1, \ldots} V_{ORM}^\infty(\boldsymbol{u}_0, \boldsymbol{u}_1, \ldots), \\ &V_{ORM}^\infty(\boldsymbol{u}_0, \boldsymbol{u}_1, \ldots) := \sum\nolimits_{k=0}^{\infty} \left\| \boldsymbol{y}_k(\boldsymbol{u}_k) - \boldsymbol{y}_k^m \right\|^2, \\ &s.t.(10.1), (10.2), (10.3), \end{aligned} \tag{10.4}$$

and $\boldsymbol{y}_k$ explicitly depends on $\boldsymbol{u}_k$ to suggest the influence of the optimization variable. The goal is to design and tune a control law $\boldsymbol{u}_k = \boldsymbol{C}(\boldsymbol{s}_k, \boldsymbol{\rho}_k)$ with $\boldsymbol{s}_k$ a regressor vector which differs from one control approach to another. When e.g. $\boldsymbol{s}_k = \boldsymbol{x}_k$, the resulting controller writes as $\boldsymbol{u}_k = \boldsymbol{C}(\boldsymbol{y}_k, ..., \boldsymbol{y}_{k-\tau}, \boldsymbol{u}_{k-1}, ..., \boldsymbol{u}_{k-\tau}, \boldsymbol{\rho}_k)$, clearly a recurrent form which can be modeled by many architectures. Especially linear ones such as the most well-known proportional-integral (PI) or proportional-integral-derivative (PID) which are familiar to control theorists and practitioners. Still, $\boldsymbol{s}_k$ will be contextualized depending on the control strategy at hand. To this end, when excited by $\boldsymbol{\rho}_k$, one should ideally obtain $\boldsymbol{y}_k = \boldsymbol{y}_k^m$, i.e. the closed-loop system's (CLS's) output (also the true system output) equals the ORM output. From control perspective, the LORM tracking/matching is in fact model reference control. From machine learning viewpoint, this is a form of supervised learning where a teacher (the ORM in our case) serves as behavioral model to the apprentice (the underlying nonlinear system (10.1) in here). This is a well formalized approach especially with robotics control.

10.2.2 The ORM Tracking with RL Design

An extended state-space model grouping the true state dynamics (10.1), the ORM dynamics (10.3) and a so-called generative reference input model (GRIM) is defined as RL needs a Markovian controlled system. Let this extended model write as

$$\boldsymbol{s}_{k+1} = \Xi(\boldsymbol{s}_k, \boldsymbol{u}_k) \iff \begin{bmatrix} \boldsymbol{x}_{k+1} \\ \boldsymbol{x}_{k+1}^m \\ \boldsymbol{\rho}_{k+1} \end{bmatrix} = \Xi\left(\begin{bmatrix} \boldsymbol{x}_k \\ \boldsymbol{x}_k^m \\ \boldsymbol{\rho}_k \end{bmatrix}, \boldsymbol{u}_k\right) = \begin{bmatrix} \boldsymbol{F}(\boldsymbol{x}_k, \boldsymbol{u}_k) \\ \boldsymbol{A}_{ORM}\boldsymbol{x}_k^m + \boldsymbol{B}_{ORM}\boldsymbol{\rho}_k \\ \Im(\boldsymbol{\rho}_k) \end{bmatrix}, \tag{10.5}$$

where $\boldsymbol{s}_k = \left[\boldsymbol{x}_k^T, \left(\boldsymbol{x}_k^m\right)^T, \boldsymbol{\rho}_k\right]^T \in \Omega_S$, with known Ω_S, since $\Omega_X, \Omega_{X^m}, \Omega_\rho$ are known, $\Xi(.,.) : \Omega_S \times \Omega_U \to \Omega_S$ is a partially unknown nonlinear map (since $\boldsymbol{F}(.)$ is partially unknown). Here, $\boldsymbol{\rho}_{k+1} = \Im(\boldsymbol{\rho}_k)$ is the user-definable GRIM. Many signals (ramps, piece-wise constants, sinusoids, but not only) are describable with this GRIM equation format.

Model-free Value Iteration (MFVIRL) is a representative RL algorithm that learns the optimal control solution (10.4), when the LORM tracking problem is defined w.r.t. the extended state-space model (10.5) instead of (10.2). And in the most extreme case, it learns it with completely unknown spate-space dynamics (10.5). The popular Q-learning algorithm, when implemented in "off-policy" "offline" style, is in fact a type of MFVIRL.

A dataset $DS = \{(\boldsymbol{s}_k^{[i]}, \boldsymbol{u}_k^{[i]}, \boldsymbol{s}_{k+1}^{[i]})\}, i = \overline{1, Z}$ of transitions (experiences/tuples) represents the information pool from which MFVIRL learns the optimal value of a so-called "Q-function". This Q-function is expressed as $Q(\boldsymbol{s}_k, \boldsymbol{u}_k)$ and extends the original cost function $V(\boldsymbol{s}_k) = \sum_{i=k}^{\infty}\gamma^{i-k}p(\boldsymbol{s}_i, \boldsymbol{u}_i)$. When the discount term

γ is set to 1 and when the penalty is $p(\boldsymbol{s}_i, \boldsymbol{u}_i) = ||\boldsymbol{y}_i - \boldsymbol{y}_i^m||^2$, it is observed that $V(\boldsymbol{s}_0) = V_{ORM}^{\infty}$ from (10.4). It is this penalty definition with serves as stimulus or incentive for teaching the controlled system's output to follow the ORM's output.

Contextually, we observe that $\boldsymbol{y}_k$ in (10.5) is obtained directly from $\boldsymbol{s}_k$ while $\boldsymbol{y}_k^m$ is $\boldsymbol{C}_{ORM}\boldsymbol{x}_k^m$, with $\boldsymbol{x}_k^m$ being part of $\boldsymbol{s}_k$. A controller (or policy) function $\boldsymbol{u}_k = \boldsymbol{C}(\boldsymbol{s}_k)$ is needed to write the Q-function Bellman equation $Q^C(\boldsymbol{s}_k, \boldsymbol{u}_k) = p(\boldsymbol{s}_k, \boldsymbol{u}_k) + \sum_{i=k+1}^{\infty} p(\boldsymbol{s}_i, \boldsymbol{C}(\boldsymbol{s}_i)) = p(\boldsymbol{s}_k, \boldsymbol{u}_k) + V^C(\boldsymbol{s}_{k+1}) = p(\boldsymbol{s}_k, \boldsymbol{u}_k) + Q^C(\boldsymbol{s}_{k+1}, \boldsymbol{C}(\boldsymbol{s}_{k+1}))$ (upper $\boldsymbol{C}$ signifies dependence on controller function $\boldsymbol{C}$). We used here the predefined value of the discount γ set to 1. However, in practical applications, the state and input domains Ω_S, Ω_U are infinitely dense despite being finite. Falling under this assumption, both the Q-function and the controller are usually modeled by (deep) neural networks (NNs) whose main objective is to interpolate (extrapolate) better. We further enumerate the MFVIRL steps in the following algorithm (named herein "*The MFVIRL Algorithm*"), where a NN models the Q-function while the controller is linearly parameterized as by $\boldsymbol{u}_k = \boldsymbol{K}^T\boldsymbol{s}_k = \boldsymbol{K}_0^T\boldsymbol{y}_k + \cdots + \boldsymbol{K}_\tau^T\boldsymbol{y}_{k-\tau} + \boldsymbol{K}_{\tau+1}^T\boldsymbol{u}_{k-1} + \cdots + \boldsymbol{K}_{2\tau+1}^T\boldsymbol{u}_{k-\tau} + \boldsymbol{K}_{2\tau+2}^T\boldsymbol{x}_k^m + \boldsymbol{K}_{2\tau+3}^T\boldsymbol{\rho}_k$ where $\boldsymbol{K}^T = [\boldsymbol{K}_0^T, \ldots, \boldsymbol{K}_\tau^T, \boldsymbol{K}_{\tau+1}^T, \ldots, \boldsymbol{K}_{2\tau+1}^T, \boldsymbol{K}_{2\tau+2}^T, \boldsymbol{K}_{2\tau+3}^T] \in \Re^{d_u \times [(\tau+1)d_y + \tau d_u + n_x + d_y]}$. In this case, the Q-function NN is formally written as $Q(\boldsymbol{s}_k, \boldsymbol{u}_k, \boldsymbol{w})$ ($\boldsymbol{w}$ are the NN's weights) and $\boldsymbol{u}_k = \boldsymbol{C}(\boldsymbol{s}_k, \boldsymbol{K}) = \boldsymbol{K}^T\boldsymbol{s}_k$.

The MFVIRL Algorithm for Reference Model Tracking

1. Select j_{max} as the maximal number of MFVIRL iterations. Select $\boldsymbol{M}(q)$ under classical model reference control rules.

2. Collect an I/O dataset of samples $\{\boldsymbol{u}_k, \boldsymbol{y}_k\}$ from (10.1) (open- or closed-loop experiment).

3. Either in the same experiment or in a following simulation, produce $\{\boldsymbol{\rho}_k, \boldsymbol{x}_k^m, \boldsymbol{y}_k^m\}$ using a prior selected GRIM $\boldsymbol{\rho}_{k+1} = \Im(\boldsymbol{\rho}_k)$. The trajectories $\boldsymbol{x}_k^m, \boldsymbol{y}_k^m$ require a state-space realization (10.3) from $\boldsymbol{M}(q)$. These steps' order is needed to ensure that Ω_Y and Ω_{Y^m} have similar span. The trajectories $\{\boldsymbol{u}_k, \boldsymbol{y}_k\}$ and $\{\boldsymbol{\rho}_k, \boldsymbol{x}_k^m, \boldsymbol{y}_k^m\}$ need to explore well their domains, (all their variables' components uniformly sample their domains). Good exploration is achievable by either sufficiently-long trajectories or by clever enhancements. Many of them exist, the most common (arguably the easiest) is using injected probing noise.

4. Choose the τ value. Reconstruct the trajectory $\{\boldsymbol{u}_k, \boldsymbol{y}_k, \boldsymbol{x}_k\}$ of the virtual state-space (10.2). Combine it with $\{\boldsymbol{\rho}_k, \boldsymbol{x}_k^m, \boldsymbol{y}_k^m\}$ in order to obtain the transition samples $\left(\boldsymbol{s}_k^{[i]}, \boldsymbol{u}_k^{[i]}, \boldsymbol{s}_{k+1}^{[i]}\right)$, $i = \overline{1, Z}$, which characterize the dynamics (10.5).

5. Choose an initial NN weight to model $Q(\boldsymbol{s}_k, \boldsymbol{u}_k, \boldsymbol{w}^0)$ ($\boldsymbol{w}^j$ are the NN's weights in the jth iteration). Choose (e.g. randomly) an initial control parameter $\boldsymbol{K}^0$ which is not necessarily admissible for the CLS.

6. At jth iteration, make $NN_{in} = \{[\boldsymbol{s}_k^{[i]^T}, \boldsymbol{u}_k^{[i]}]^T\}$ and $NN_{out} = \{p(\boldsymbol{s}_k^{[i]}, \boldsymbol{u}_k^{[i]}) + Q(\boldsymbol{s}_{k+1}^{[i]}, \boldsymbol{K}^{j-1^T}\boldsymbol{s}_{k+1}^{[i]}, \boldsymbol{w}^{j-1})\}$, for all $i = \overline{1, Z}$, as I/O NN training patterns, respectively. By selecting NN architecture and training hyper-parameters, the NN is trained based on these patterns. This is the VI-specific Q-function

evaluation step, it equivalates with minimizing the MSSE cost as $w^j = \underset{w}{argmin}\frac{1}{Z}\sum_{i=1}^{Z}\left(Q\left(s_k^{[i]}, u_k^{[i]}, w\right) - p\left(s_k^{[i]}, u_k^{[i]}\right) - Q(s_{k+1}^{[i]}, K^{j-1^T} s_{k+1}^{[i]}, w^{j-1})\right)^2$ using e.g. well-known backprop gradient-based search techniques.

7. Over each $s_k^{[i]}, i = \overline{1, Z}$, compute $u_k^{[i]^*} = \underset{u}{\text{argmin}}\, Q\left(s_k^{[i]}, u, w^j\right)$. One feasible solution for a bounded but infinitely-dense domain is to enumerate a finite set of equally-spaced values for u. This is efficiently implementable on personal computers, for MIMO systems with up to three inputs ($d_u = 3$) even for Z up to the order of 10,000 samples. A linear equation system

$$\begin{bmatrix} s_k^{[1]^T} \\ \cdots \\ s_k^{[Z]^T} \end{bmatrix} K^j = \begin{bmatrix} u_k^{[1]^{*T}} \\ \cdots \\ u_k^{[Z]^{*T}} \end{bmatrix} \Longleftrightarrow AK^j = b, \quad (10.6)$$

follows, which is overdetermined and solvable as $K^j = \left(A^T A\right)^{-1} A^T b$.

8. If MFVIRL iteration index j reaches its maximal allowed value j_{max}, or there is no improvement in the Q-function NN parameter, i.e. $\|w^j - w^{j-1}\| < \upsilon$ for some threshold υ, then exit the algorithm; otherwise increment j and return to Step 6.

The stopping condition from Step 8 could have been based on $\|K^j - K^{j-1}\|$ as well. When working with NNs as approximators for the function Q, the training is always noisy to some extent, although the controller gain K derived based on this Q-function may stabilize quicker. Therefore, a larger threshold υ would be used for the stopping condition in terms of w's evolution.

MFVIRL offline learning mode is in fact an "experience replay"-based concept since in the current implementation, all collected transition samples simultaneously participate in the learning process. Learning convergence has been analyzed in recent works, most often under ideal exploration setting [35]. Therefore, it resembles the more popular Q-learning algorithm from artificial intelligence however, without online but with offline exploration, and without using replay buffers. Instead, all samples are used altogether during learning.

10.2.3 Learning ORM Tracking with VSFRT Control

The VSFRT feedback control utilizes—as its name suggests—the virtual state information, therefore it uses the state-space model (10.2) at its core, but with another strategy for offline virtual reference input obtention [2]. VSFRT extends and generalizes the more popular nonlinear multivariable VRFT control method in such that the control is not a direct function of the output feedback error ($u_k = C(e_k = \rho_k - y_k)$) but a more generic functional $u_k = C\left(x_k, \rho_k\right)$. Defining the extended regressor $s_k = \left[x_k^T, \rho_k^T\right]^T$, then the VSFRT control looks like $u_k = C(s_k)$. We point out that in this form, both terms ρ_k, y_k are embedded inside s_k. The starting input for VSFRT is a collection of I/O samples $\{\boldsymbol{u}_k, \boldsymbol{y}_k\}$ measured from the original system (10.1) in either open- or closed-loop with prior stabilizing controller used for exploratory data collection. Then VSFRT subsequently generates a virtual trajectory $\{\boldsymbol{u}_k, \boldsymbol{y}_k, \tilde{\boldsymbol{x}}_k\}$ of (10.2), in offline mode since (10.2) is just an alias representation of the true and unknown state-space corresponding to (10.1), but in a different unknown coordinate transformation. To denote offline computation procedure, we use the "~" symbol on top of the variable. The offline-generated so-called virtual reference is subsequently obtained as $\tilde{\boldsymbol{\rho}}_k = \boldsymbol{M}^{-1}(q)\boldsymbol{y}_k$. This virtual reference in not employed as an actual CLS input in any kind of interaction with the system (10.1), it is only an intermediate signal (hence the name "virtual") and its obtention stems from the idea that $\boldsymbol{y}_k$ should simultaneously be the output from the CLS and from the ORM. Let us construct $\left\{\boldsymbol{s}_k = \left[\tilde{\boldsymbol{x}}_k^T, \tilde{\boldsymbol{\rho}}_k^T\right]^T\right\}, k = \overline{1, Z}$ as an extended state regressor database. The VRFT rationale tries to solve (10.4) by transforming it to a controller identification problem which makes that the controller depending on $\boldsymbol{s}_k$ (and generically being nonlinearly dependent on parameter $\boldsymbol{\pi}$) to obtain $\boldsymbol{u}_k$ at the output, where $\boldsymbol{u}_k$ is the one obtained in the initial measurement experiment. The controller identification writes as

$$\boldsymbol{\pi}^* = \underset{\boldsymbol{\pi}}{\operatorname{argmin}} V_{VR}^N(\boldsymbol{\pi}), \; V_{VR}^N(\boldsymbol{\pi}) = \frac{1}{Z}\sum\nolimits_{k=1}^{Z} ||\boldsymbol{u}_k - \boldsymbol{C}(\boldsymbol{s}_k, \boldsymbol{\pi})||^2. \tag{10.7}$$

Different from the MFVIRL, the state $\boldsymbol{s}_k$ in VSFRT is independent on the state $\boldsymbol{x}_k^m$ from the ORM, as it is regarded problematic in the underlying regression in (10.7) [2]. To this end, the ORM could be known only by its PTM $\boldsymbol{M}(q)$ while its state-space is unnecessary. Again, different from MFVIRL, VSFRT does not need a GRIM, owing to its different computing principle of $\boldsymbol{\rho}_k$. Context-dependent discussions are required. E.g., as $\boldsymbol{M}(q)$ is usually a low-pass character, it determines $\boldsymbol{M}^{-1}(q)$ acting high-pass. For noisy outputs $\boldsymbol{y}_k$, prefiltering with a low-pass filter is then critical. This prefilter is slightly differently perceived than the classical pre-filter $\boldsymbol{L}(q)$ in VRFT, not being applied to all components of $\boldsymbol{s}_k$ nor to $\boldsymbol{u}_k$. In fact, for richly parameterized controllers (such as NNs), the VRFT filter could be omitted [2, 31, 33, 35, 47]. Then, even for non-causal $\boldsymbol{M}^{-1}(q)$, offline implementation is feasible.

The VSFRT steps are enumerated in the following algorithm, when used with a linear controller function ($\boldsymbol{\pi} = \boldsymbol{K}$) where $\boldsymbol{u}_k = \boldsymbol{K}^T\boldsymbol{s}_k = \boldsymbol{K}_0^T y_k + \cdots + \boldsymbol{K}_\tau^T y_{k-\tau} + \boldsymbol{K}_{\tau+1}^T\boldsymbol{u}_{k-1} + \cdots + \boldsymbol{K}_{2\tau+1}^T\boldsymbol{u}_{k-\tau} + \boldsymbol{K}_{2\tau+2}^T\boldsymbol{\rho}_k$, where $\boldsymbol{K}^T = [\boldsymbol{K}_0^T, \ldots, \boldsymbol{K}_\tau^T, \boldsymbol{K}_{\tau+1}^T, \ldots, \boldsymbol{K}_{2\tau+1}^T, \boldsymbol{K}_{2\tau+2}^T] \in \Re^{(\tau+1)d_y+\tau d_u+d_y}$ (different from MFVIRL linear state feedback, the ORM's $\boldsymbol{x}_k^m$ is missing). The motivation behind a linear parameterization is that, for many slightly nonlinear systems, a linear state-feedback ensures satisfactory and yet robust performance. Extension to the nonlinear parameterized controller case (e.g. with NNs approximator) is trivial and was validated on numerous occasions.

The VSFRT Algorithm for Reference Model Tracking

1. Select the LORM PTM $\boldsymbol{M}(q)$ following model reference control guidelines. Collect an I/O trajectory $\{\boldsymbol{u}_k, \boldsymbol{y}_k\}$ from (10.1) (either in open- or in closed-loop) where the persistency of excitation (PE) condition for $\boldsymbol{u}_k$ needs to be fulfilled to ensure that $\boldsymbol{y}_k$ encodes all the underlying system's modes (a similar exploration condition was required by MFVIRL). Injected probing noise is the simplest way of ensuring the PE condition.

2. Choose the value of τ and construct the virtual trajectory $\{\boldsymbol{u}_k, \boldsymbol{y}_k, \tilde{\boldsymbol{x}}_k\}$ characterizing the dynamics from (10.2).

3. Obtain the virtual reference $\tilde{\boldsymbol{\rho}}_k$ with the noncausal filtering operation $\tilde{\boldsymbol{\rho}}_k = \boldsymbol{M}^{-1}(q)\boldsymbol{y}_k$. Preprocessing of $\boldsymbol{y}_k$ using low pass filtering may desirable whenever it is affected by high-frequency content such as noise.

4. Obtain $\left\{ \boldsymbol{s}_k = \left[\tilde{\boldsymbol{x}}_k^T, \tilde{\boldsymbol{\rho}}_k^T \right]^T \right\}, k = \overline{1, Z}$.

5. Write the linear equations system explicated in matrix form

$$\begin{bmatrix} \boldsymbol{s}_1^T \\ \dots \\ \boldsymbol{s}_Z^T \end{bmatrix} \boldsymbol{K} = \begin{bmatrix} \boldsymbol{u}_1^T \\ \dots \\ \boldsymbol{u}_Z^T \end{bmatrix} \iff \boldsymbol{A}\boldsymbol{K} = \boldsymbol{b}, \tag{10.8}$$

we solve the overdetermined system as $\boldsymbol{K} = \left(\boldsymbol{A}^T\boldsymbol{A}\right)^{-1}\boldsymbol{A}^T\boldsymbol{b}$. This actually minimizes (10.7) w.r.t. $\boldsymbol{\pi} = \boldsymbol{K}$.

Note that the VSFRT algorithm is not an iterative approach like MFVIRL, it is "one-shot". The closed-loop system was proven stable under VSFRT control, as it has been analyzed in recent literature [2]. At least from conceptual and computational perspective, VSFRT has an advantage over MFVIRL: it does not keep an approximation for the value function, it is also not iterative in the control learning process.

In what follows, application of MFVIRL and VSFRT is presented on three inexpensive, lab-scale systems.

10.3 ORM Tracking for the Electrical Braking System

10.3.1 The Electrical Braking System Description

A diagram of the EBS is depicted in Fig. 10.1. It is a rheostatic brake emulator whose objective is to keep a constant voltage V_{gen} when the source voltage V_{source} is varying. This means controlling the current flow through the blue line in the circuit. It is achieved by gradually opening or closing the transistor by changing its base voltage such that the current flowing through the collector-emitter junction varies. A pseudo rheostatic braking effect is achieved by keeping a constant V_{source} and by

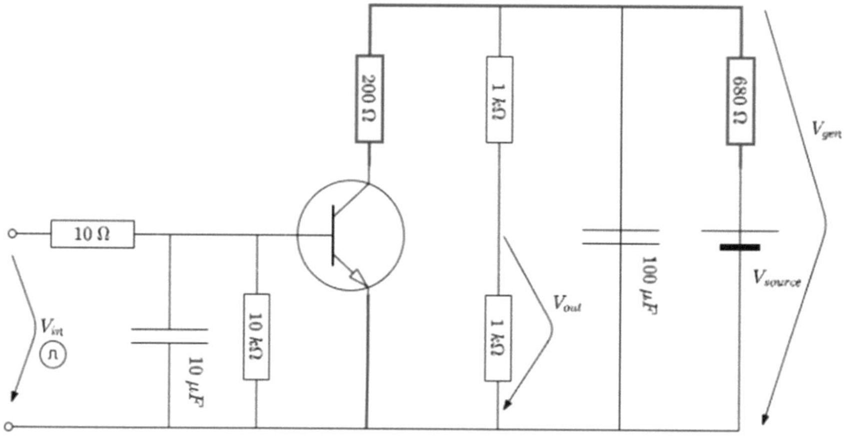

Fig. 10.1 Diagram of the electrical braking system (EBS) [50]

changing the voltage V_{gen} to a desired level, since this is cheaper to systematically reproduce than changing V_{source}.

Apart from the electrical characteristics of the resistors and capacitors, a value of $V_{source} = 9V$ is used, the BJT NPN transistor module is a TIP31C whose base voltage complies with the voltage level output from a PWM port of an Arduino®. Due compatible microcontroller-based interface. The measured voltage V_{out} across the divider brings V_{gen} to a measurable domain of the ADC port of the Arduino interface. The voltage $V_{in} \in [0; 5]V$ is changed by the PWM duty cycle as control input u_k, hence the variation of $y_k = V_{gen} \in [2; 6.7]V$ to the desired level is obtained. Arduino serves as a communication interface with MATLAB/Simulink at a sampling period of $T_s = 0.05s$. Such a system's cost can be as low as 10$, it does not require special components but ones that are widely available and cheap. The system is inspired by [50].

10.3.2 The I/O Data Collection Experiment Settings

The I/O collected samples is the data pool from which both MFVIRL and VSFRT will learn control. The EBS is first included in a closed-loop with an integral-type controller with transfer function $0.1/s$. Then the reference input to the resulted closed-loop system is a uniform random number (URN) with amplitude in $[2.2; 6]V$, with sample period of $5s$. Another URN with amplitude in $[-0.1; 0.1]$ and sample period $2T_s$ is added on the controller output as probing noise.

For generating the ORM input-state-output data, ρ_k is an URN with amplitude in $[2.2; 6]V$ and sample period of $6.5s$. It therefore models a piecewise-constant signal with GRIM $\rho_{k+1} = \rho_k$ except for the switching sampling instants. The trajectories x_k^m, y_k^m result based on the ORM dynamics given first in continuous-time domain transfer function $M(s) = 1/(0.4s + 1)$ and afterwards discretized with zero-order hold to $M(q)$. The discretized ORM is finally transformed to the state-space controllable form

$$x_{k+1}^m = 0.8825 x_k^m + 0.1175 \rho_k,$$
$$y_k^m = x_k^m. \tag{10.9}$$

A total number of 2000 samples of each signal from the previously mentioned are collected. We emphasize that ρ_k, x_k^m, y_k^m could be generated in real time or offline as well. It is important to overlap the domain of y_k^m with that of y_k. This sort of calibration is better performed offline.

10.3.3 The MFVIRL Application Settings

For the MFVIRL algorithm, $\tau = 3$ is set, leading to $\boldsymbol{s}_k = \left[y_k, y_{k-1}, y_{k-2}, y_{k-3}, u_{k-1}, u_{k-2}, u_{k-3}, x_k^m, \rho_k\right]^T \in \mathfrak{R}^9$. The transition samples $\left(\boldsymbol{s}_k^{[i]}, \boldsymbol{u}_k^{[i]}, \boldsymbol{s}_{k+1}^{[i]}\right)$, $i = \overline{1, 1982}$ are built by excluding the samples from the switching instants where $\rho_{k+1} \neq \rho_k$. The Q-function is modeled by a deep NN with ten inputs (the size of $[\boldsymbol{s}_k^T, u_k]$), two hidden layers with 100 and 50 Rectified Linear Units (ReLU) activation functions and with a single output neuron with linear activation. The training data is prepared at each iteration of the MFVIRL algorithm, according to the steps. A 90–10% training/validation ratio is used, with uniformly random extracted samples. Scaled conjugated gradient method with random NN weights initialized, maximum 10 consecutive increases in the MSSE cost early stops the training to avoid overfitting, maximum 500 epochs are used, the batch size in each epoch is maximal ($Z = 1982$). $\boldsymbol{K}^0 = \boldsymbol{0}$ initially, and Q-function NN minimization over each $\boldsymbol{s}_k^{[i]}$ is done by enumerating 22 uniformly samples values in the domain [0; 1] of the control input u_k, i.e. $\{0, 0.0476, 0.0952, 0.1429, \ldots, 1\}$. The minimizing arguments for the Q value NN over the input u_k are used in the linear equation system to be solved for the improved controller. 200 MFVIRL iteration are executed before the algorithm stops with $\boldsymbol{K}^{200} = [-0.4449, 0.1029, 0.1216, -0.0803, 0.0436, 0.2420, 0.2153, 0.1813, 0.1769]^T$. The resulted ORM tracking control performance with this optimal MFVIRL controller is shown in Fig. 10.2.

10.3.4 The VSFRT Application Settings

For the VSFRT algorithm, the value of $\tau = 3$ is set and used to construct the virtual trajectory $\{u_k, y_k, \tilde{\boldsymbol{x}}_k\}$ of (10.2). Obviously, the same ORM $M(q)$ is selected since the same goal is pursued by VSFRT as for MFVIRL. To obtain the virtual reference

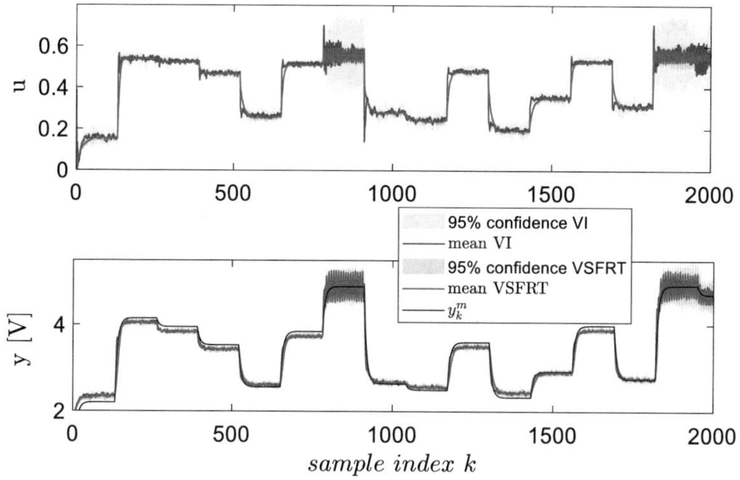

Fig. 10.2 The statistical ORM tracking results: five runs for each of the MFVIRL and VSFRT controllers, with confidence intervals and mean trajectory

input as $\tilde{\rho}_k = M^{-1}(q)y_k$, we note that y_k is noisy therefore we first low-pass filter it through $\frac{0.1}{1-0.9q^{-1}}$. Only afterwards, noncausal offline filtering the output renders $\tilde{\rho}_k$. The extended regressor state is finally built as $\left\{ \boldsymbol{s}_k = \left[\tilde{\boldsymbol{x}}_k^T, \tilde{\rho}_k^T \right]^T \right\}, k = \overline{1, 1997}$. This time, there is no Markovian assumption to be obeyed by the virtual reference input. The solution to the overdetermined linear equations system renders $\boldsymbol{K} = [-0.0855, 0.0584, -0.0666, 0.0614, 0.9134, -0.4072, 0.4488, 0.0373]^T \in \Re^8$ which points out the missing x_k^m. The resulted ORM tracking control performance with this optimal VSFRT controller is shown in Fig. 10.2.

To correctly evaluate the ORM tracking performance, the value $fit_{\%} = 100\Big(1 - \sum_{i=1}^N (y^k - y_k^m)^2 / (\sum_{i=1}^N (y^k - \overline{y}_k)^2)\Big)$ was measured, where the mean value is defined $\overline{y}_k = 1/N \sum_{i=1}^N y_k$, for $N = 2000$. This metric has good practical interpretation and well correlates with the value V_{LRMO}^{∞} in (10.4).

Figure 10.2 and Table 10.1 reflect good ORM tracking performance from both MFVIRL and VSFRT controllers. Also, the linear controller cannot uniformly satisfy the ORM tracking across all operating points (more oscillations around the higher setpoints is visible with both controllers). This is a structural limitation, in cases such as this a nonlinear controller could better account for the operating point. The missing integral component from the controllers is also visible in terms of small steady-state errors at various setpoint levels.

Table 10.1 Fit score $fit_{\%}$ with the MFVIRL and VSFRT learned controllers, obtained for the EBS

Trial	$fit_{\%}$	
	MFVIRL	VSFRT
1	82.23	81.88
2	82.95	82.07
3	82.80	81.92
4	82.51	81.64
5	82.67	81.75
Mean	**82.63**	81.85
StDev σ	0.2772	**0.1645**

10.4 ORM Tracking for the Active Temperature Control System (ATCS)

10.4.1 The Active Temperature Control System Description

The ATCS is shown in Fig. 10.3 and it consists of a heating element and a cooling element. A power supply of $V_{source} = 12$ V capable of 2 amps (A) maximum current will power either the heater which is a TIP31C NPN transistor module whose base voltage level is Arduino compatible, or the cooler which is a DC motor fan of 0.5 A and 7000 rpm at maximum voltage. A LM35DZ analog temperature sensor is set to touch the transistor body and thermal paste is applied between the two, to improve the heat transfer coefficient. An aluminum heatsink is mounted on the transistor (which is driven in ON–OFF switching mode), to better dissipate heat. The DC motor fan blows room temperature air on the transistor and sensor module and helps at cooling

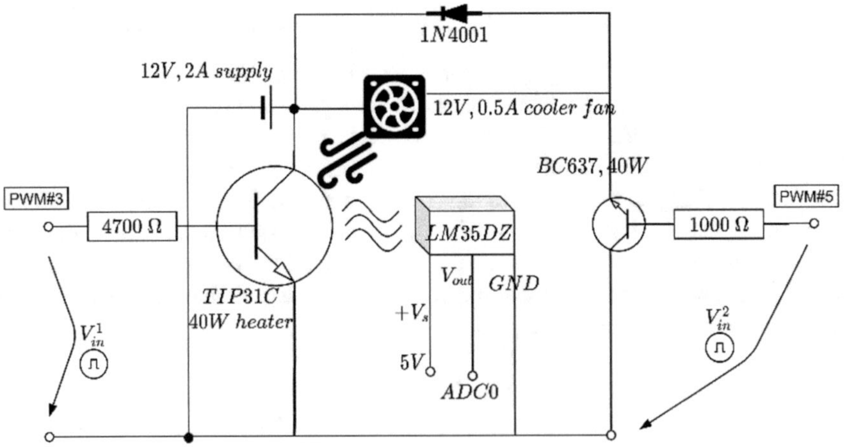

Fig. 10.3 Diagram of the active temperature control system (ATCS)

the transistor ensemble, which is more advantageous than the natural heat dissipation process having longer time constant. The DC fan control logic uses another transistor driven in ON–OFF switching mode, together with a 1N4001 flyback diode. The sample period for the ATCS is $T_s = 20s$ since it is a slow process.

There is a single control input $u \in [-1; 1]$ representing signed PWM duty cycle, that will be employed to alternatively drive the heater of the cooler. The activation logic is

$$\begin{cases} V_{in}^1 = \max(\min(0.2 + |u|, 1), 0) \times 5V, \ V_{in}^2 = 0V, \ \ when\, u \geq 0 \\ V_{in}^1 = 0V, \ V_{in}^2 = \max(\min(0.15 + |u|, 1), 0) \times 5V, \ \ when\, u < 0 \end{cases} \tag{10.10}$$

where the values 0.2 and 0.15 applied in V_{in}^1 and V_{in}^2, respectively, are used to compensate the dead-zones in the heater and cooler, respectively. This is because, the transistor is open below 1 V and drives no current (therefore generates no heat), while the DC motor fan is spinless for its transistor base voltage below 0.75 V. All voltages are saturated inside $[0; 5]V$ using the max, min operators. The process output is $y = V_{out}$ measured on the analog-to-digital Arduino port. This voltage value is in fact a straightforward normalized Celsius temperature, i.e. y is measured in $^\circ C/100$, which will be used for control feedback purposes. This system's cost can be as low as 15\$, it does not require special components. In our case, the 12 V supply was obtained from a DC-DC buck-boost converter about 4\$ in value. The system has extended functionality over similarly existent [51].

10.4.2 The I/O Data Collection Experiment Settings

The I/O data samples collected from the ATCS represents the data pool from which both MFVIRL and VSFRT will learn control. The ATCS is interacted with in open-loop, with the input u_k being modeled as a piecewise constant signal with URN amplitudes in $[-0.3; 0.8]$, constant for 2200 s. For exploration enhancement, another additive URN with amplitude in $[-0.5; 0.5]$ is added to the base one as a probing noise, this time being piecewise-constant for 100 s.

For generating the ORM input-state-output data, ρ_k is an URN with amplitude in $[0.3; 0.98]\,^\circ C/100$ and sample period of 2100 s. It therefore models a piecewise-constant signal with GRIM $\rho_{k+1} = \rho_k$ except for the switching sampling instants. The trajectories $\boldsymbol{x}_k^m$, y_k^m result based on the ORM dynamics given first in continuous-time domain transfer function $M(s) = 1/(500s + 1)$ and afterwards discretized with zero-order hold to $M(q)$. The discretized ORM $M(q)$ is finally transformed to the state-space controllable form

$$\begin{aligned} x_{k+1}^m &= 0.9608 x_k^m + 0.0392 \rho_k, \\ y_k^m &= x_k^m. \end{aligned} \tag{10.11}$$

A total number of 4000 samples of each signal are collected. From all the collected evidence, the system has very slow dynamics, the data collection and testing prove to be time consuming activities. This would be even longer if passive cooling would have been used, instead of using the active cooling solution based on the fan.

10.4.3 The MFVIRL Application Settings for ATCS

For the MFVIRL algorithm, $\tau = 2$ is set, leading to $s_k = \left[y_k, y_{k-1}, y_{k-2}, u_{k-1}, u_{k-2}, x_k^m, \rho_k\right]^T \in \mathfrak{R}^7$. The transition samples $\left(s_k^{[i]}, u_k^{[i]}, s_{k+1}^{[i]}\right)$, $i = \overline{1, 3962}$ are built by excluding the samples from the switching instants where $\rho_{k+1} \neq \rho_k$. The Q-function is modeled by a deep NN with ten inputs (the size of $[s_k^T, u_k]$), two hidden layers with 100 and 50 Rectified Linear Units (ReLU) activation functions and with a single output neuron with linear activation. The training data is prepared at each iteration of the MFVIRL algorithm, according to the steps. A 90–10% training/validation ratio is used, with uniformly random extracted samples. Scaled conjugated gradient method with random NN weights initialized, maximum 10 consecutive increases in the MSSE cost early stops the training to avoid overfitting, maximum 500 epochs are used, the batch size in each epoch is maximal ($Z = 3962$). $\boldsymbol{K}^0 = \boldsymbol{0}$ initially, and Q-function NN minimization over each $s_k^{[i]}$ is done by enumerating 22 uniformly samples values in the control input u_k domain $[-1; 1]$, i.e. $\{-1, -0.9048, \ldots, 0.9048, 1\}$. The minimizing arguments for the Q value NN over the input u_k are used in the linear equation system to be solved for the improved controller. 100 MFVIRL iteration are executed before the algorithm stops with $\boldsymbol{K}^{100} = [-4.3133, 1.1058, 0.4347, -0.0915, 0.1707, 2.3851, 0.7131]^T$. The resulted ORM tracking control performance with this optimal MFVIRL controller is shown in Fig. 10.4.

10.4.4 The VSFRT Application Settings for ATCS

For the VSFRT algorithm, the value of $\tau = 2$ is set and used to construct the virtual trajectory $\{u_k, y_k, \tilde{\boldsymbol{x}}_k\}$ of (10.2). Obviously, the same ORM $M(q)$ is selected since the same goal is pursued by VSFRT as for MFVIRL. To obtain the virtual reference input as $\tilde{\rho}_k = M^{-1}(q)y_k$, we note that y_k is noisy therefore we first low-pass filter it through $\frac{0.5}{1-0.5q^{-1}}$. Only afterwards, noncausal offline filtering the output renders $\tilde{\rho}_k$. The extended regressor state is finally built as $\left\{s_k = \left[\tilde{\boldsymbol{x}}_k^T, \tilde{\rho}_k\right]^T\right\}, k = \overline{1, 3998}$. Again, no Markovian assumption about the virtual reference input is needed. The overdetermined linear equations system solves to renders $\boldsymbol{K} = [-9.4666, -1.8942, 10.5299, 0.4443, 0.4441, 0.8854]^T \in \mathfrak{R}^6$ which

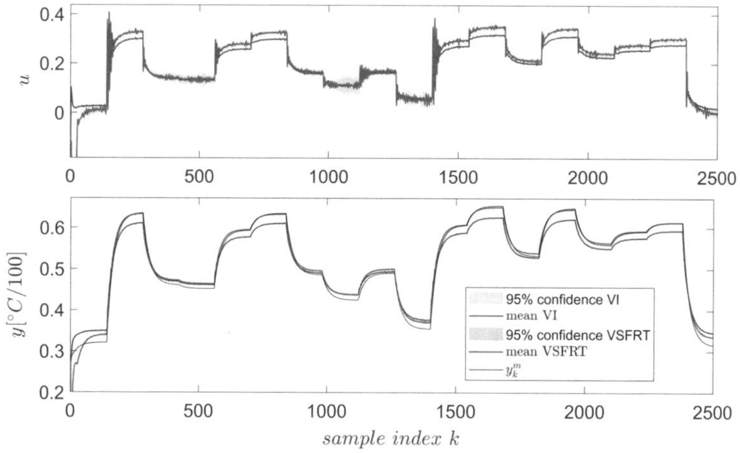

Fig. 10.4 The statistical ORM tracking results: five runs for each of the MFVIRL (blue) and VSFRT (red) controllers, with confidence intervals and mean trajectory (in shaded colored areas)

points out the missing x_k^m. The resulted ORM tracking control performance with this optimal VSFRT controller is shown in Fig. 10.4.

For fairer ORM tracking evaluation, the scoring index $fit_{\%} = 100\left(1 - \sum_{i=1}^{N}\left(y^k - y_k^m\right)^2 / (\sum_{i=1}^{N}\left(y^k - \overline{y}_k\right)^2)\right)$ was again measured, where $\overline{y}_k = 1/N\sum_{i=1}^{N} y_k$ defines the mean value of the controlled output, for $N = 2500$.

Figure 10.4 and Table 10.2 show a satisfactory ORM tracking performance from both MFVIRL and VSFRT control. The integral component missing from the controllers leads to small steady-state errors at various setpoint levels. In this case study, the VSFRT control is systematically better than the MFVIRL control. This is dependent also on the collected I/O data quality in terms of exploration. In the presented application, the linear controller structure does not display severe structural limitation, being able to handle various operating points to a satisfactory level.

Table 10.2 Fit score $fit_{\%}$ with the MFVIRL and VSFRT learned controllers, obtained for the ATCS

Trial	$fit_{\%}$	
	MFVIRL	VSFRT
1	79.7302	87.3714
2	79.7424	87.3804
3	79.7397	87.4074
4	79.7361	87.3664
5	79.7212	87.3885
Mean	79.7339	**87.3828**
StDev σ	**0.0085**	0.0161

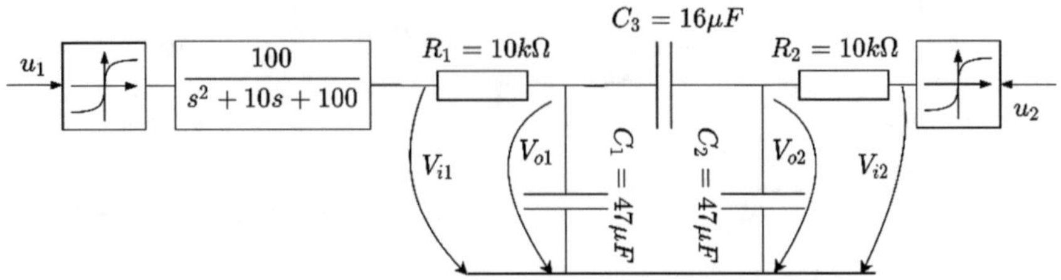

Fig. 10.5 Diagram of the voltage control electrical system and its hardware realization [52]

10.5 ORM Tracking for the Voltage Control Electrical System (VCES)

10.5.1 The Multivariable Voltage Control Electrical System (VCES) Description

The hybrid software-electrical VCES shown in Fig. 10.5 has its two control inputs $u_1 \in [0; 1]$, $u_2 \in [0; 1]$ as PWM duty cycles for setting the input voltages $V_{i1}, V_{i2} \in [0; 5]V$ in the electrical system. The two inputs are first passed through a nonlinear hyperbolic tangent static function (known as "tanh") running on the PC while the first input is also filtered through a resonant second order normalized transfer function, to add additional dynamics (also processed on the PC). We aim at bringing the voltages across the capacitors C_1 and C_2 respectively, to the desired levels. That is, $y_1 = V_{o1}[V]$, $y_2 = V_{o2}[V]$ are the controlled outputs. Another capacitor C_3 allows coupled behavior between the two control channels. We offset the controlled inputs to $[-0.5; 0.5]$ while the controlled outputs are also offset to $[-2.5; 2.5]V$ by subtracting $2.5V$ from the voltages obtained after converting them using the Arduino analog-to-digital converter. This way, the operation is symmetric around the zero leveled inputs and outputs. The sample period for this system is $T_s = 0.1s$ and it is sufficiently small to capture the system transient dynamic and to interface the system with MATLAB/Simulink. The 2×2 multivariable system has $u_{k,1}, u_{k,2}$ as discrete-time inputs and $y_{k,1}, y_{k,2}$ as discrete-time outputs. The VCES costs can be as low as 5–7$, it uses widely available components.

10.5.2 The I/O Data Collection Experiment Settings for the VCES

The I/O data samples are collected from the VCES and they provide the sufficient information pool for learning MFVIRL and VSFRT control. Open-loop interaction is sought with piecewise constant $u_{k,1}, u_{k,2}$ being both URNs within $[-0.5; 0.5]$

with their sample periods of $3T_s$ and $2T_s$, respectively. Different seeds are used for generating them.

For ORM input-state-output data generation, the ORM is first selected as the 2-by-2 transfer matrix $M(q) = [\frac{e^{-0.1s}}{s+1} \quad 0; 0 \quad \frac{e^{-0.1s}}{s+1}]$. Its size reflects the two control channels and the two controlled outputs. Additionally, a unit delay is included in the transfer functions, since it a unit time delay has been observed in the system from each input to each output. Moreover, decoupling is imposed by the diagonal ORM. The piecewise constant $\rho_{k,1}$, $\rho_{k,2}$ are both URNs with their amplitudes within $[-1; 1]$ and their sample periods of $5.5s$ and $5s$, respectively. To reconstruct the input-state-output ORM data $\boldsymbol{x}_k^m$, $\boldsymbol{y}_k^m$ by feeding the ORM with $\rho_{k,1}$, $\rho_{k,2}$, a canonical controllable state-space transform is used [52]

$$\begin{cases} \begin{bmatrix} s_{k+1,1}^m \\ \varpi_{k+1,1} \\ s_{k+1,2}^m \\ \varpi_{k+1,2} \end{bmatrix} = \begin{bmatrix} 0.9048 & 0.0952 & 0 & 0 \\ 0 & 0 & 0 & 0 \\ 0 & 0 & 0.9048 & 0.0952 \\ 0 & 0 & 0 & 0 \end{bmatrix} \begin{bmatrix} s_{k,1}^m \\ \varpi_{k,1} \\ s_{k,2}^m \\ \varpi_{k,2} \end{bmatrix} + \begin{bmatrix} 0 & 0 \\ 1 & 0 \\ 0 & 0 \\ 0 & 1 \end{bmatrix} \begin{bmatrix} \rho_{k,1} \\ \rho_{k,2} \end{bmatrix}. \\ \boldsymbol{y}_k^m = [y_{k,1}^m y_{k,2}^m]^T = [1, 0, 1, 0] \times [s_{k,1}^m, \varpi_{k,1}, s_{k,2}^m, \varpi_{k,2}]^T. \end{cases} \tag{10.12}$$

Five thousand samples of each signal are collected in the exploratory phase. We stress again that the ORM signals could be generated offline, not necessarily in real-time with the VCES running experiment. It is important to calibrate the domain of the ORM outputs $\boldsymbol{y}_k^m$ such that it overlaps with the output domain of the VCES.

10.5.3 The MFVIRL Application Settings for VCES

For the MFVIRL algorithm, $\tau = 3$ is set, leading to $s_k = \left[y_{k,1}, y_{k-1,1}, y_{k-2,1}, y_{k-3,1}, y_{k,2}, y_{k-1,2}, y_{k-2,2}, y_{k-3,2}, u_{k-1,1}, u_{k-2,1}, u_{k-3,1}, u_{k-1,2}, u_{k-2,2}, u_{k-3,2}, \boldsymbol{x}_k^{mT}, \boldsymbol{\rho}_k^T\right]^T \in \mathfrak{R}^{20}$. The transition samples $\left(\boldsymbol{s}_k^{[i]}, \boldsymbol{u}_k^{[i]}, \boldsymbol{s}_{k+1}^{[i]}\right)$, $i = \overline{1, 4802}$ are built by excluding the samples from the switching instants where $\boldsymbol{\rho}_{k+1} \neq \boldsymbol{\rho}_k$. The Q-function is modeled by a deep NN with 22 inputs (the size of $\left[\boldsymbol{s}_k^T, \boldsymbol{u}_k^T\right] \in \mathfrak{R}^{22}$), two hidden layers with 100 and 50 Rectified Linear Units (ReLU) activation functions and with a single output neuron with linear activation. The training data is prepared at each iteration of the MFVIRL algorithm, according to the steps. A 90–10% training/validation ratio is used, with uniformly random extracted samples. Scaled conjugated gradient method with random NN weights initialized, maximum 10 consecutive increases in the MSSE cost early stops the training to avoid overfitting, maximum 500 epochs are used, the batch size in each epoch is maximal ($Z = 4802$). $\boldsymbol{K}^0 = \boldsymbol{0}$ initially, and Q-function NN minimization over each $\boldsymbol{s}_k^{[i]}$ is done by enumerating 22 uniformly samples values in the control input $\boldsymbol{u}_k$ domain $[-1; 1] \times [-1; 1]$, i.e. $\{-1, -0.9048, \ldots, 0.9048, 1\} \times$

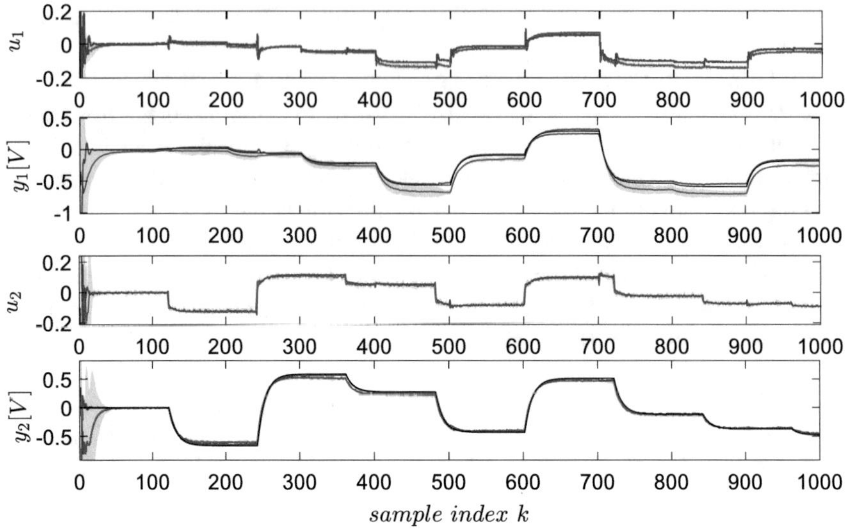

Fig. 10.6 The statistical ORM tracking results: five runs for each of the MFVIRL (blue) and VSFRT (red) controllers, with confidence intervals and mean trajectory (in shaded areas). $y^m_{k,1}$, $y^m_{k,2}$ are in black

$\{-1, -0.9048, \ldots, 0.9048, 1\}$. The minimizing arguments for the Q value NN over the input $\boldsymbol{u}_k$ are used in the linear equation system to be solved for the improved controller. 100 MFVIRL iteration are executed before the algorithm stops with

$$\boldsymbol{K}^{100} = \begin{matrix} [-0.5041, -0.2243, -0.1172, 0.1538, 0.1857, 0.2349, 0.0729, -0.2963, \\ -0.3981, -0.3905, -0.1484, -0.1044, -0.1676, -0.2404, 0.8379, 0.0354, \\ -0.0856, -0.0970, 0.1420, 0.0514; \\ \\ -0.1253, -0.1119, 0.2299, -0.1141, -0.9592, -0.0587, 0.1157, \\ -0.0096, -0.0517, 0.0057, 0.078, -0.0485, 0.0967, -0.0134, 0.0842, \\ -0.0307, 0.9467, 0.1336, 0.0654, -0.0147]. \end{matrix} \tag{10.13}$$

The resulted ORM tracking control performance with this optimal MFVIRL controller is shown in Fig. 10.6.

10.5.4 The VSFRT Application Settings for VCES

For the VSFRT algorithm, the value of $\tau = 3$ is set and used to construct the virtual trajectory $\{\boldsymbol{u}_k, \boldsymbol{y}_k, \tilde{\boldsymbol{x}}_k\}$ of (10.2). Obviously, the same ORM $\boldsymbol{M}(q)$ is selected since the same goal is pursued by VSFRT as for MFVIRL. The virtual reference input as $\tilde{\boldsymbol{\rho}}_k = \boldsymbol{M}^{-1}(q)\boldsymbol{y}_k$, no prefilter is used. With diagonal $\boldsymbol{M}(q)$, each virtual reference input is calculated from its corresponding output. The extended regressor state is finally built as $\left\{\boldsymbol{s}_k = \left[\tilde{\boldsymbol{x}}_k^T, \tilde{\boldsymbol{\rho}}_k^T\right]^T\right\}, k = \overline{1,4983}$. Again, since no Markovian assumption about the virtual reference input is needed, all samples are used for learning. The overdetermined linear equations system solves to, renders

$$\boldsymbol{K}^{100} = \begin{matrix} [0.1052, 0.3646, -0.1036, -0.1951, 0.0993, -0.1307, -0.0363, 0.1082, \\ -0.3222, -0.0864, -0.3256, 0.0013, -0.0336, 0.0590, 0.2070, -0.0412; \\ 0.2298, 0.2724, -0.2658, -0.1759, -0.0522, -0.6335, -1.0223, 1.4151, \\ 0.0726, -0.0824, -0.1258, 0.7524, -0.1632, 1.3671, -0.0337, 0.0886]. \end{matrix} \tag{10.14}$$

The resulted ORM tracking control performance with this optimal VSFRT controller is shown in Fig. 10.6.

For fairer evaluation of the ORM tracking, the scoring index $fit_{\%} = 100\left(1 - \left[\sum_{i=50}^{1000}\left(y_{k,1} - y_{k,1}^m\right)^2 + \sum_{i=50}^{1000}\left(y_{k,2} - y_{k,2}^m\right)^2\right] / \left[\sum_{i=50}^{1000}\left(y_{k,1} - \overline{y}_{k,1}\right)^2 + \sum_{i=50}^{1000}\left(y_{k,2} - \overline{y}_{k,2}\right)^2\right]\right)$ was again measured, where $\overline{y}_{k,1} = 1/951 \times \sum_{i=50}^{1000} y_{k,1}, \overline{y}_{k,2} = 1/951 \times \sum_{i=50}^{1000} y_{k,2}$ defines the mean values of the controlled outputs. In this case, the statistics are calculated beginning with the 50th sample (5 s) after the test scenario starts, to allow sufficient time for the resulted control system to bring the outputs near zero.

Figure 10.6 and Table 10.3 show a satisfactory ORM tracking performance from both MFVIRL and VSFRT control. As with the previous case studies, the missing integral component in the controllers leads to small steady-state errors at various setpoint levels. In this case study, the VSFRT control is inferior with respect to the MFVIRL control, the latter also being more robust.

Table 10.3 Fit score $fit_{\%}$ with the MFVIRL and VSFRT learned controllers, obtained for the VCES

Trial	$fit_{\%}$	
	MFVIRL	VSFRT
1	93.0618	83.8657
2	93.3125	84.3834
3	93.3098	84.6671
4	93.2730	84.4172
5	93.2779	85.9227
Mean	**93.2470**	84.6512
StDev σ	**0.1051**	0.7682

10.6 Comments and Conclusions

The two proposed intelligent learning control approaches for ORM tracking have been validated on three, affordable lab scale systems which are of high importance to practical home/industrial automation. Although linear controllers were employed for nonlinear systems, it can be considered as a form of general nonlinear control [53–57] for which particular attention is dedicated to tracking applications [58–64]. The learning principle proposed here is model-free, avoiding the use of system dynamics knowledge. It is based on I/O data, and it implements simple, linear controller structures with both MFVIRL and VSFRT. The systems' diversity, both in phenomenology, nonlinearity and multidimensionality, demonstrate the potential that these two techniques have for both industrial applications and educational purposes.

In prior work [2] it was observed that VSFRT always produced a superior performance, even in the cases when poor exploration was performed when collecting the I/O data. This is somewhat different than the observed behavior in the current work where MFVIRL has shown in some cases that it can produce a superior control solution. Most evidence points to the fact that the collected I/O data quality in terms of exploration, is to blame for the alternating superior performance of the two analyzed intelligent control techniques. It is of importance and stringent need for future research direction, to define proper exploration quality metrics and correlate them with learned ORM tracking performance. This could lead to inductive analysis to select the experiment design in both exploration quality and safety, models' suitable parameterizations, data volumes needed for guaranteed learning, etc.

Another issue is with the proposed controller parameterization, which is constraining to some extent and therefore limiting the performance. However, using the same linear controller parameterization for both techniques offers a fair perspective regarding their performance. The issue of selecting appropriate parameterizations for value function and controller approximations in MFVIRL and for the controller parameterization in VSFRT are open issues which trade-off many implementation aspects: learning convergence rate, data volume requirements, maximum achievable performance in the case of general nonlinear systems and implementation costs in terms of computational complexity. Finding a good trade-off is exclusively based on designer experience nowadays, but the near future expects automated tools to help with the design.

Finally, validating the proposed techniques on even more various systems would prove their learning capacity. Comparisons with simple classical controllers such as (but not limited to) PI/PIDs, will also benchmark their performance versus complexity versus data volume requirements.

Acknowledgements This work was supported by a grant of the Ministry of Research, Innovation and Digitization, CNCS/CCCDI—UEFISCDI, project number PN-III-P1-1.1-TE-2019-1089, within PNCDI III.

The first author would like to thank Eng. Cristian Schlezinger, Dr. Robert Antal and Eng. Catalin Petrică, for the help and recommendations in designing the EBS.

References

1. Radac, M.B., Lala, T.: Robust control of unknown observable nonlinear systems solved as a zero-sum game. IEEE Access **8**, 214153–214165 (2020). https://doi.org/10.1109/ACCESS.2020.3040185
2. Radac, M.B., Borlea, A.I.: Virtual state feedback reference tuning and value iteration reinforcement learning for unknown observable systems control. Energies **14**, 1006 (2021). https://doi.org/10.3390/en14041006
3. Mnih, V., Kavukcuoglu, K., Silver, D., et al.: Human-level control through deep reinforcement learning. Nature **518**, 529–533 (2015). https://doi.org/10.1038/nature14236
4. De Bruin, T., Kober, J., Tuyls, K., Babuska, R.: Integrating state representation learning into deep reinforcement learning. IEEE Robot. Autom. Lett. **3**, 1394–1401 (2018). https://doi.org/10.1109/LRA.2018.2800101
5. Lewis, F.L., Vamvoudakis, K.G.: Reinforcement learning for partially observable dynamic processes: adaptive dynamic programming using measured output data. IEEE Trans. Syst. Man Cybern. Part B Cybern. **41**, 14–25 (2011). https://doi.org/10.1109/TSMCB.2010.2043839
6. Radac, M.B., Precup, R.E., Petriu, E.M.: Model-free primitive-based iterative learning control approach to trajectory tracking of mimo systems with experimental validation. IEEE Trans. Neural Netw. Learn. Syst. **26**, 2925–2938 (2015). https://doi.org/10.1109/TNNLS.2015.2460258
7. Radac, M.B., Precup, R.E.: Three-level hierarchical model-free learning approach to trajectory tracking control. Eng. Appl. Artif. Intell. **55**, 103–118 (2016). https://doi.org/10.1016/j.engappai.2016.06.009
8. Wu, B., Gupta, J.K., Kochenderfer, M.: Model primitives for hierarchical lifelong reinforcement learning. Auton. Agent Multi Agent Syst. **34**, 1–28 (2020). https://doi.org/10.1007/s10458-020-09451-0
9. Li, J., Li, Z., Li, X., et al.: Skill learning strategy based on dynamic motion primitives for human-robot cooperative manipulation. IEEE Trans. Cogn. Dev. Syst. **13**, 105–117 (2021). https://doi.org/10.1109/TCDS.2020.3021762
10. Kim, Y.L., Ahn, K.H., Song, J.B.: Reinforcement learning based on movement primitives for contact tasks. Robot. Comput. Integr. Manuf. **62**, 101863 (2020). https://doi.org/10.1016/j.rcim.2019.101863
11. Camci, E., Kayacan, E.: Learning motion primitives for planning swift maneuvers of quadrotor. Auton. Robots **43**, 1733–1745 (2019). https://doi.org/10.1007/s10514-019-09831-w
12. Yang, C., Chen, C., He, W., et al.: Robot learning system based on adaptive neural control and dynamic movement primitives. IEEE Trans. Neural Netw. Learn. Syst. **30**, 777–787 (2019). https://doi.org/10.1109/TNNLS.2018.2852711
13. Werbos, P.J.: A menu of designs for reinforcement learning over time. In: Miller, W.T., Sutton, R.S.¸ Werbos, P.J. (eds.) Neural Networks for Control, pp. 67–95.. MIT Press, Cambridge, MA (1990)
14. Lewis, F.L., Vrabie, D.: Reinforcement learning and adaptive dynamic programming for feedback control. IEEE Circuits Syst. Mag. **9**, 32–50 (2009). https://doi.org/10.1109/MCAS.2009.933854
15. Murray, J.J., Cox, C.J., Lendaris, G.G., Saeks, R.: Adaptive dynamic programming. IEEE Trans. Syst. Man Cybern. Part C Appl. Rev. **32**, 140–153 (2002). https://doi.org/10.1109/TSMCC.2002.801727
16. Wang, F.Y., Zhang, H., Liu, D.: Adaptive dynamic programming: an introduction. IEEE Comput. Intell. Mag. **4**, 39–47 (2009). https://doi.org/10.1109/MCI.2009.932261
17. Fu, H., Chen, X., Wang, W., Wu, M.: MRAC for unknown discrete-time nonlinear systems based on supervised neural dynamic programming. Neurocomputing **384**, 130–141 (2020). https://doi.org/10.1016/j.neucom.2019.12.023
18. Wang, W., Chen, X., Fu, H., Wu, M.: Data-driven adaptive dynamic programming for partially observable nonzero-sum games via Q-learning method. Int. J. Syst. Sci. **50**, 1338–1352 (2019). https://doi.org/10.1080/00207721.2019.1599463

19. Perrusquia, A., Yu, W.: Neural H2 control using continuous-time reinforcement learning. IEEE Trans. Cybern. 1–10 (2020). https://doi.org/10.1109/TCYB.2020.3028988
20. Sardarmehni, T., Heydari, A.: Sub-optimal switching in anti-lock brake systems using approximate dynamic programming. IET Control Theory Appl. **13**, 1413–1424 (2019). https://doi.org/10.1049/iet-cta.2018.5428
21. Martinez-Piazuelo, J., Ochoa, D.E., Quijano, N., Giraldo, L.F.: A multi-critic reinforcement learning method: an application to multi-tank water systems. IEEE Access **8**, 173227–173238 (2020). https://doi.org/10.1109/ACCESS.2020.3025194
22. Liu, Y., Zhang, H., Yu, R., Xing, Z.: H∞ tracking control of discrete-time system with delays via data-based adaptive dynamic programming. IEEE Trans. Syst. Man Cybern. Syst. **50**, 4078–4085 (2020). https://doi.org/10.1109/TSMC.2019.2946397
23. Buşoniu, L., de Bruin, T., Tolić, D., et al.: Reinforcement learning for control: performance, stability, and deep approximators. Annu. Rev. Control **46**, 8–28 (2018). https://doi.org/10.1016/j.arcontrol.2018.09.005
24. Na, J., Lv, Y., Zhang, K., Zhao, J.: Adaptive identifier-critic-based optimal tracking control for nonlinear systems with experimental validation. IEEE Trans. Syst. Man Cybern. Syst. 1–14 (2020). https://doi.org/10.1109/tsmc.2020.3003224
25. Huang, M., Liu, C., He, X., et al.: Reinforcement learning-based control for nonlinear discrete-time systems with unknown control directions and control constraints. Neurocomputing **402**, 50–65 (2020). https://doi.org/10.1016/j.neucom.2020.03.061
26. Treesatayapun, C.: Knowledge-based reinforcement learning controller with fuzzy-rule network: experimental validation. Neural Comput. Appl. **32**, 9761–9775 (2020). https://doi.org/10.1007/s00521-019-04509-x
27. Campi, M.C., Lecchini, A., Savaresi, S.M.: Virtual reference feedback tuning: a direct method for the design of feedback controllers. Automatica **38**, 1337–1346 (2002). https://doi.org/10.1016/S0005-1098(02)00032-8
28. Formentin, S., Savaresi, S.M., Del Re, L.: Non-iterative direct data-driven controller tuning for multivariable systems: theory and application. IET Control Theory Appl. **6**, 1250–1257 (2012). https://doi.org/10.1049/iet-cta.2011.0204
29. Campestrini, L., Eckhard, D., Gevers, M., Bazanella, A.S.: Virtual reference feedback tuning for non-minimum phase plants. Automatica **47**, 1778–1784 (2011). https://doi.org/10.1016/j.automatica.2011.04.002
30. Eckhard, D., Campestrini, L., Christ Boeira, E.: Virtual disturbance feedback tuning. IFAC J. Syst. Control **3**, 23–29 (2018). https://doi.org/10.1016/j.ifacsc.2018.01.003
31. Yan, P., Liu, D., Wang, D., Ma, H.: Data-driven controller design for general MIMO nonlinear systems via virtual reference feedback tuning and neural networks. Neurocomputing **171**, 815–825 (2016). https://doi.org/10.1016/j.neucom.2015.07.017
32. Campi, M.C., Savaresi, S.M.: Direct nonlinear control design: the virtual reference feedback tuning (VRFT) approach. IEEE Trans. Automat. Control **51**, 14–27 (2006). https://doi.org/10.1109/TAC.2005.861689
33. Esparza, A., Sala, A., Albertos, P.: Neural networks in virtual reference tuning. Eng. Appl. Artif. Intell. **24**, 983–995 (2011). https://doi.org/10.1016/j.engappai.2011.04.003
34. Radac, M.B., Precup, R.E.: Data-driven model-free slip control of anti-lock braking systems using reinforcement Q-learning. Neurocomputing **275**, 317–329 (2018). https://doi.org/10.1016/j.neucom.2017.08.036
35. Radac, M.B., Precup, R.E.: Data-driven model-free tracking reinforcement learning control with VRFT-based adaptive actor-critic. Appl. Sci. **9**, 1807 (2019). https://doi.org/10.3390/app9091807
36. Sjöberg, J., Gutman, P.O., Agarwal, M., Bax, M.: Nonlinear controller tuning based on a sequence of identifications of linearized time-varying models. Control Eng. Pract. **17**, 311–321 (2009). https://doi.org/10.1016/j.conengprac.2008.08.001
37. Wang, I.J., Spall, J.C.: Stochastic optimisation with inequality constraints using simultaneous perturbations and penalty functions. Int. J. Control **81**, 1232–1238 (2008). https://doi.org/10.1080/00207170701611123

38. Mišković, L., Karimi, A., Bonvin, D., Gevers, M.: Correlation-based tuning of decoupling multivariable controllers. Automatica **43**, 1481–1494 (2007). https://doi.org/10.1016/j.automatica.2007.02.006
39. Safonov, M.G., Tsao, T.C.: The unfalsified control concept and learning. IEEE Trans. Automat. Control **42**, 843–847 (1997). https://doi.org/10.1109/9.587340
40. Krstić, M.: Performance improvement and limitations in extremum seeking control. Syst. Control Lett. **39**, 313–326 (2000). https://doi.org/10.1016/S0167-6911(99)00111-5
41. Bolder, J., Kleinendorst, S., Oomen, T.: Data-driven multivariable ILC: enhanced performance by eliminating L and Q filters. Int. J. Robust Nonlinear Control **28**, 3728–3751 (2018). https://doi.org/10.1002/rnc.3611
42. Chi, R., Hou, Z., Jin, S., Huang, B.: An improved data-driven point-to-point ILC using additional on-line control inputs with experimental verification. IEEE Trans. Syst. Man Cybern. Syst. **49**, 687–696 (2019). https://doi.org/10.1109/TSMC.2017.2693397
43. Zhang, J., Meng, D.: Convergence analysis of saturated iterative learning control systems with locally Lipchitz nonlinearities. IEEE Trans. Neural Netw. Learn. Syst. **31**, 4025–4035 (2020). https://doi.org/10.1109/TNNLS.2019.2951752
44. Li, X., Chen, S.L., Teo, C.S., Tan, K.K.: Data-based tuning of reduced-order inverse model in both disturbance observer and feedforward with application to tray indexing. IEEE Trans. Ind. Electron. **64**, 5492–5501 (2017). https://doi.org/10.1109/TIE.2017.2674623
45. Hui, Y., Chi, R., Huang, B., et al.: Observer-based sampled-data model-free adaptive control for continuous-time nonlinear nonaffine systems with input rate constraints. IEEE Trans. Syst. Man Cybern. Syst. 1–10 (2020). https://doi.org/10.1109/tsmc.2020.2982491
46. Fliess, M., Join, C.: An alternative to proportional-integral and proportional-integral-derivative regulators: intelligent proportional-derivative regulators. Int. J. Robust Nonlinear Control (2021). https://doi.org/10.1002/rnc.5657
47. Radac, M.B., Precup, R.E.: Data-driven MIMO model-free reference tracking control with nonlinear state-feedback and fractional order controllers. Appl. Soft Comput. J. **73**, 992–1003 (2018). https://doi.org/10.1016/j.asoc.2018.09.035
48. Radac, M.B., Precup, R.E., Roman, R.C.: Data-driven model reference control of MIMO vertical tank systems with model-free VRFT and Q-learning. ISA Trans. **73**, 227–238 (2018). https://doi.org/10.1016/j.isatra.2018.01.014
49. Radac, M.B., Precup, R.E., Roman, R.C.: Model-free control performance improvement using virtual reference feedback tuning and reinforcement Q-learning. Int. J. Syst. Sci. **48**, 1071–1083 (2017). https://doi.org/10.1080/00207721.2016.1236423
50. Petrica, C.: Voltage control in a rheostatical brake simulator. Thesis, Politehnica University of Timisoara, Romania, B.Sc (2021)
51. Hedengren, J.D.: Advanced Temperature Control. https://apmonitor.com/do/index.php/Main/AdvancedTemperatureControl. Accessed 12 Sept 2021
52. Lala, T., Radac, M.B.: Learning to extrapolate an optimal tracking control behavior towards new tracking tasks in a hierarchical primitive-based framework. In: Proceedings of IEEE 2021 29th Mediterranean Conference on Control and Automation (MED), June 22–25, 2021. Bari, Italy, pp. 421–427 (2021)
53. Cao, S., Sun, L., Jiang, J., Zuo, Z.: Reinforcement learning-based fixed-time trajectory tracking control for uncertain robotic manipulators with input saturation. IEEE Trans. Neural Netw. Learn. Syst. (2021). https://doi.org/10.1109/TNNLS.2021.3116713
54. Dong, F., Jin, D., Zhao, X., Han, J., Lu, W.: A non-cooperative game approach to the robust control design for a class of fuzzy dynamical systems. ISA Trans. (2021). https://doi.org/10.1016/j.isatra.2021.06.031
55. Chai, Y., Luo, J., Ma, W.: Data-driven game-based control of microsatellites for attitude takeover of target spacecraft with disturbance. ISA Trans. (2021). https://doi.org/10.1016/j.isatra.2021.02.037
56. Dogru, O., Velswamy, K., Huang, B.: Actor-critic reinforcement learning and application in developing computer-vision-based interface tracking. Engineering (2021). https://doi.org/10.1016/j.eng.2021.04.027

57. Li, H., Wang, Y., Pang, M.: Disturbance compensation based model-free adaptive tracking control for nonlinear systems with unknown disturbance. Asian J. Control **23**, 708–717 (2021). https://doi.org/10.1002/asjc.2230
58. Lee, W., Jeoung, H., Park, D., Kim, T., Lee, H., Kim, N.: A real-time intelligent energy management strategy for hybrid electric vehicles using reinforcement learning. IEEE Access **9**, 72759–72768 (2021). https://doi.org/10.1109/ACCESS.2021.3079903
59. Moreno-Valenzuela, J., Montoya-Cháirez, J., Santibáñez, V.: Robust trajectory tracking control of an underactuated control moment gyroscope via neural network-based feedback linearization. Neurocomputing **403**, 314–324 (2020). https://doi.org/10.1016/j.neucom.2020.04.019
60. Fei, Y., Shi, P., Lim, C.C.: Robust and collision-free formation control of multiagent systems with limited information. IEEE Trans. Neural Netw. Learn. Syst. (2021). https://doi.org/10.1109/TNNLS.2021.3112679
61. Meng, X., Yu, H., Xu, T., Wu, H.: Disturbance observer and L2-gain-based state error feedback linearization control for the quadruple tank liquid-level system. Energies **13**, 5500 (2020). https://doi.org/10.3390/en13205500
62. Mohammadzadeh, A., Vafaie, R.H.: A deep learned fuzzy control for inertial sensing: micro electro mechanical systems. Appl. Soft. Comput. **109**, 10759 (2021). https://doi.org/10.1016/j.asoc.2021.107597
63. Zhao, H., Peng, L., Yu, H.: Model-free adaptive consensus tracking control for unknown nonlinear multi-agent systems with sensor saturation. Int. J. Robust Nonlinear Control **31**, 6473–6491 (2021). https://doi.org/10.1002/rnc.5630
64. Zhao, J., Na, J., Gao, G.: Robust tracking control of uncertain nonlinear systems with adaptive dynamic programming. Neurocomputing (2021). https://doi.org/10.1016/j.neucom.2021.10.081

Printed in the United States
by Baker & Taylor Publisher Services